MW00561761

BACTERIA
ARCHAEA
SYNTHETICA
EUCARYA
SYNTHETIC
BIOLOGY
A PRIMER

Erratum for Chapter 9

This chapter was written by Filippa Lentzos, Caitlin Cockerton, Susanna Finlay, R. Alexander Hamilton, Joy Yueyue Zhang and Nikolas Rose, of the Department of Social Science, Health & Medicine, King's College London.

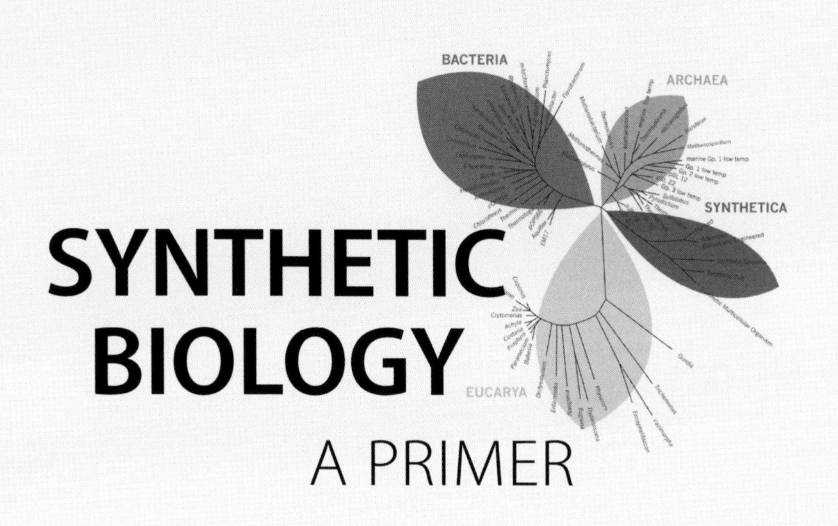

SYNTHETIC BIOLOGY

A PRIMER

Geoff Baldwin
Travis Bayer
Robert Dickinson
Tom Ellis
Paul S Freemont
Richard I Kitney
Karen Polizzi
Guy-Bart Stan

Imperial College, UK

ICP

Imperial College Press

Published by

Imperial College Press
57 Shelton Street
Covent Garden
London WC2H 9HE

Distributed by

World Scientific Publishing Co. Pte. Ltd.
5 Toh Tuck Link, Singapore 596224
USA office: 27 Warren Street, Suite 401-402, Hackensack, NJ 07601
UK office: 57 Shelton Street, Covent Garden, London WC2H 9HE

British Library Cataloguing-in-Publication Data
A catalogue record for this book is available from the British Library.

SYNTHETIC BIOLOGY — A PRIMER

ISBN-13 978-1-84816-862-6
ISBN-10 1-84816-862-4
ISBN-13 978-1-84816-863-3 (pbk)
ISBN-10 1-84816-863-2 (pbk)

Typeset by Stallion Press
Email: enquiries@stallionpress.com

Printed by FuIsland Offset Printing (S) Pte Ltd Singapore

Front cover — The Synthetic Kingdom by Daisy Ginsberg

How will we classify what is natural or unnatural when life is built from scratch? The Tree of Life is always changing, ever since we first created it. Now, synthetic biology is turning to the living kingdoms for its materials library. No more petrochemicals: instead, pick a feature from an existing organism, locate its DNA and insert into a biological chassis. Engineered life will compute, produce energy, clean up pollution, kill pathogens and even do the housework.

These synthetic organisms are no different from other life forms, except that we invented them. We'll simply have to insert an extra branch to the Tree of Life to classify them. Perhaps the Synthetic Kingdom is part of our new nature?

The Synthetic Kingdom mirrors synthetic biology's ideology: it's a future fashioned by engineering logic, a rationalisation of the complexity of living systems, an engineering solution to an engineering problem. But it also puts our designs back into the complexity of nature rather than separating us from them.

Contents

List of Contributors

Editors and Contributors:

Paul Freemont (p.freemont@imperial.ac.uk)

Paul Freemont is co-principal investigator of the Molecular Structure and Function laboratory, Chair of Protein Crystallography, Head of the Division of Molecular Biosciences and co-Director of the EPSRC Centre for Synthetic Biology and Innovation at Imperial College London. He joined Imperial in 2000 as Director of the Centre for Structural Biology after leaving the Cancer Research UK London Institute, where he established experimental structural biology research in 1988, following a postdoctoral fellowship at Yale University working with Professor Tom Steitz. Professor Freemont's interdisciplinary research interests have focused on understanding the molecular mechanisms of disease-linked proteins. He has also developed statistical approaches for the spatial analysis of nuclear architecture in normal and cancer cells. More recently he has developed research interests in synthetic biology, concentrating on biosensors for pathogenic organisms and part/device characterisation protocols. He co-founded the EPSRC Centre for Synthetic Biology and Innovation with Professor Richard Kitney. The Centre is the first of its kind in the UK and aims to develop an engineering framework and new technology platforms to enable synthetic biology research in areas of bioenergy, biosensors, biomaterials and metabolic engineering. He and Professor Richard Kitney have been responsible for six highly successful imperial college iGEM teams.

Richard Kitney (r.kitney@imperial.ac.uk)

Richard Kitney is Professor of BioMedical Systems Engineering, Chairman of the Institute of Systems and Synthetic Biology and co-Director of the EPSRC Centre for Synthetic Biology and Innovation. He was the co-Chair of the joint Inquiry by the Royal Academy of Engineering and the Academy of Medical Sciences on Systems Biology. The report of the Inquiry — *Systems Biology: a vision for engineering and medicine* — was published in February 2007. He also was Chair of the Royal Academy of Engineering Inquiry into Synthetic Biology — *Synthetic Biology: scope, applications and implications* was published in May 2009. He is also a member of the Royal Society Working Party on Synthetic Biology. He has worked extensively in the United States and has been a Visiting Professor at MIT since 1991. He is a co-Director of the Imperial College–MIT International Consortium for Medical Information Technology. Professor Kitney is now working extensively in synthetic biology and is heading Imperial College's initiative in this area with Professor Paul Freemont. They have been responsible for six highly successful Imperial College iGEM teams.

Contributors:

Geoff Baldwin (g.baldwin@imperial.ac.uk)

Geoff Baldwin joined Imperial College in 2000 as a BBSRC David Phillips Fellow and is now a Reader in Biochemistry within the Division of Molecular Biosciences and Centre for Synthetic Biology and Innovation. A chemistry graduate from the University of East Anglia, he moved into molecular biology and biochemistry during his PhD at the University of Sheffield on DNA modifying enzymes. He followed his PhD studies with a postdoctoral fellowship at the University of Bristol, Department of Biochemistry, working with Professor Steve Halford FRS. Dr Baldwin still continues to work across interdisciplinary boundaries and in recent years his quantitative approach to studying biological systems at the molecular level has found a new outlet in synthetic biology. His research includes elucidating DNA repair mechanisms, developing methods for *in vivo* directed evolution, DNA assembly and part characterisation.

Travis Bayer (t.bayer@imperial.ac.uk)

Travis Bayer joined Imperial College in 2010 as an investigator at the Centre for Synthetic Biology and Innovation and a lecturer in the Division of Molecular Biosciences. Dr Bayer obtained his undergraduate degree at the University of Texas at Austin and his PhD at Caltech, working with Professor Christina Smolke. He followed his PhD studies with a postdoctoral fellowship at UCSF, working with Professor Chris Voigt on metabolic pathway engineering using synthetic biology. Dr Bayer's laboratory at Imperial aims to understand the structure, function and evolution of complex metabolic and regulatory systems with a goal to develop biological technologies to enhance global health and sustainability. As exemplars he is working on metabolic pathways for renewable liquid fuels, materials and pharmaceutical precursors.

Robert Dickinson (robert.dickinson@imperial.ac.uk)

Robert Dickinson is a senior lecturer in the Department of Bioengineering, Imperial College. He graduated with a degree in physics from Cambridge University, and then obtained a PhD in biophysics from the University of London, in ultrasound signal processing. Dr Dickinson has extensive experience in medical imaging, in both hospital and industrial environments. He worked on MRI coil development and system integration at Picker International Ltd, and ultrasound imaging in a small start-up company, where he developed a sub-1mm intravascular ultrasound imaging catheter for imaging coronary arteries. He has substantial experience in bio-compatibility and other issues of invasive medical devices, together with commercialisation and IP transfer. He has filed over 12 patents, and has CE-marked a number of medical devices. He has worked with Emcision Ltd on their range of electrosurgical devices.

Tom Ellis (t.ellis@imperial.ac.uk)

Tom Ellis joined Imperial College in 2010 as an investigator at the Centre for Synthetic Biology and Innovation and a lecturer in the Department of Bioengineering. Dr Ellis's undergraduate degree was from Oxford University, and his PhD from the University of Cambridge examined the use of DNA-binding drugs as synthetic gene expression regulators. Two years after his PhD and following work at a drug-discovery biotech, Dr Ellis joined one of the founding groups of synthetic biology, working under Professor Jim Collins at Boston University. Here Dr Ellis worked on engineering biology through transcriptional regulation and model-guided design of gene regulatory networks. His group at Imperial College now carries out research on predictable construction of synthetic chromosomes, regulatory pathways and industrially relevant biosensors, working in both traditional model organisms and extremophiles.

Karen Polizzi (k.polizzi@imperial.ac.uk)

Karen Polizzi joined Imperial College in September 2008 as an RCUK Fellow in Biopharmaceutical Processing within the Division of Molecular Biosciences. Dr Polizzi obtained her PhD in chemical and biomolecular engineering at the Georgia Institute of Technology, working with Professor Andreas Bommarius on protein engineering of biocatalysts. Dr Polizzi's laboratory is interested in applying synthetic biology to bioprocess engineering with a particular focus on the production of therapeutic proteins and the use of *in vivo* biosensors for developing and controlling metabolic pathways related to protein production and glycosylation. She is also interested in enzymatic processing related to the production of therapeutics.

Guy-Bart Stan (g.stan@imperial.ac.uk)

Dr Guy-Bart Stan joined Imperial College in 2010 as a principal investigator at the Centre for Synthetic Biology and Innovation and a lecturer in the Department of Bioengineering. Dr Stan is currently leading the Control Engineering Synthetic Biology group within the Centre. He obtained his PhD in applied sciences on nonlinear dynamical systems and control from the University of Liege, Belgium, in 2005. His PhD thesis dealt with the global analysis and synthesis of limit cycle oscillations in networks of interconnected nonlinear dynamical systems, and with the global synchronisation of oscillations in such networks. Dr Stan followed his PhD as a research associate in the Control Group at the Department of Engineering, University of Cambridge, before joining Imperial College. His main research interests are in the areas of nonlinear dynamical systems design, analysis and control, synthetic biology and systems biology.

Preface

Synthetic biology is an exciting and rapidly evolving new research field with the potential to transform our fundamental understanding of biological systems as well as our ability to manipulate them for human benefit. The origins of synthetic biology can be traced back to the early 1960s when developments in our understanding of the genetic code in the form of DNA led to the 'central dogma' of molecular biology where DNA encodes RNA, which encodes protein macromolecules. This powerful dogma, combined with techniques first developed in the 1970s that allowed the manipulation, transfer and cloning of DNA, has underpinned much of our understanding of molecular biosciences and cell biology and has quickly led to the genome revolution which we are currently experiencing. This rapid growth in molecular and genomic understanding was enabled by the parallel developments of new technologies and capabilities, including increasing computing power and the establishment of sophisticated information systems, as well as the ability to rapidly sequence DNA. All of these components culminated in 2000 with a landmark moment in human scientific understanding, namely the complete genetic sequence of a human being. In 2011, we now have complete genome sequences for nearly every major class of organism on the planet, and for the first time we have complete lists of the basic components (open reading frames that would encode proteins) that constitute living systems, accessible from any web browser in the world, including from a mobile phone. This rapid accumulation of biological information is correlated with the increasing interdisciplinary nature of biological research. It is perhaps not surprising to see physicists, chemists and computing scientists attracted to answering fundamental questions of living processes. Engineering science (principally systems analysis and signal processing) has been applied for many years to the analysis of biological systems in systems biology. The new discipline of synthetic biology now extends the application of engineering science to synthesis. One unique aspect of synthetic biology is therefore the fusion of engineers and engineering principles with molecular biologists and systems biology thinking. One of the driving forces in synthetic biology is the desire to make biological design and implementation easier and more predictable in biotechnology applications like bioenergy and biomanufacturing. The conceptual engineering framework that underpins synthetic biology research (modularisation, standardisation and characterisation) translates the techniques used in engineering design (such as specifications, mathematical modelling and prototyping) to the design of biological systems. The idea for this textbook came from the establishment of Master's training programmes at Imperial College London in the Centre for Synthetic Biology and Innovation. As our new students began their interdisciplinary training in synthetic biology we realised that there wasn't a source textbook that we could recommend, which has now led to this text.

Synthetic Biology — A Primer aims to give a broad overview of the emerging field of synthetic biology and the foundational concepts on which it is built. It will be primarily of interest to final year undergraduates, postgraduates and established researchers who are interested in learning more about this exciting new field. The book introduces readers to fundamental concepts in molecular biology and engineering and then explores the two major themes for synthetic biology, namely 'bottom-up' and 'top-down' engineering approaches. 'Top-down' engineering utilises a conceptual framework of engineering and systematic design to build new biological systems by integrating robustly characterised biological parts into an existing system through the use of extensive mathematical modelling. The 'bottom-up' approach involves the design and building of synthetic protocells using basic chemical and biochemical building blocks from scratch. Exemplars of cutting-edge applications designed using synthetic biology principles are presented, including the production of novel biofuels from renewable feedstocks, microbial synthesis of pharmaceuticals and fine chemicals, and the design and implementation of biosensors to detect infections and environmental waste. The book also uses the International Genetically Engineered Machine (iGEM) competition to illustrate the power of synthetic biology as an innovative research and training science. Finally, the primer includes a chapter on the ethical, legal and societal issues surrounding synthetic biology, illustrating the legitimate integration of social sciences in synthetic biology research. This is a joint contribution from our social science colleagues in the BIOS Centre at King's College London, including postgraduate students and senior researchers, and thus all the authors are cited at the start of the chapter.

We hope that students and professional researchers will find this text useful as an introduction to synthetic biology. If readers become inspired, as we have, by the potential of synthetic biology to tackle the pressing problems of our planet via biological solutions combined with a responsible overview, then we will be delighted. The enthusiasm, energy, excitement and commitment shown by our iGEM team, junior researchers, colleagues in the Royal College of Art and the London School of Economics and Political Science and King's College London, suggest that synthetic biology has a bright future.

Paul Freemont and Richard Kitney, Imperial College London, summer 2011.

CHAPTER 1

Introduction to Biology

1.1 Introduction

This introduction is intended to provide a brief overview of the basic principles that lie at the heart of biology. It is intended for engineers and physical scientists who do not have any knowledge of biology but who are interested in synthetic biology. Having an understanding of the basic principles will aid the physical scientist and engineer to model biological systems and also to communicate with biologists. There are of course many excellent textbooks dedicated to detailed descriptions of the molecular processes that occur within biology, and readers who wish to extend their knowledge beyond this brief survey are encouraged to seek out more details in the relevant texts.

There are two basic concepts that you need to understand if you wish to be able to engineer biological systems: how information flows in biological systems and how this information flow is controlled. With an understanding of these concepts one can, in principle, apply engineering principles to the design and building of new biological systems: what we call synthetic biology.

Biology is, of course, highly complex and there are important differences that distinguish it from other engineering disciplines. Firstly, biology is not programmed on a printed circuit board, so interactions cannot be programmed by their physical position; rather interactions are based on interactions between molecules that occur in the complex milieu of the cell. Secondly, biology is subject to natural selection, so that modifications which are deleterious to the cell will be selected against and competed out of the population. These evolutionary pressures are not applicable when building an aircraft, and so new definitions of robustness are relevant to biology. Other concepts such as complexity and emergent behaviour may be familiar to engineers, but one must be aware of how they can arise in biology and what their effects may be.

1.2 Information Storage in Biology

1.2.1 *DNA structure*

It is widely known by even non-specialists that DNA (deoxyribonucleic acid) is the molecule that stores the information within biological systems. The double-helical structure of the

DNA molecule has also entered the wider consciousness and provides a totemic symbol of the genomic age in which we live. To fully appreciate how this molecule functions as a dynamic information store, we must explore its structure in more detail.

The double-helical structure of DNA arises because it is composed of two separate strands. Each strand is a long molecule formed by the linking together of the individual building blocks, the nucleotides. Each nucleotide has three important parts: the phosphate, the deoxyribose sugar and the base (Fig. 1.1). The phosphate and sugar always remain the same, but there are four different bases: A, G, C and T (adenine, guanine, cytosine and thymine). The phosphate and the sugar link together to form the backbone of each DNA strand, while the bases decorate the backbone and provide variability. It is the sequence of these nucleotides that provides the basis of information coding in DNA.

A key feature of DNA, which leads to the double-helix structure, is how the two strands interact. Each of the four bases can only pair with one of the other bases; A can only pair with T, and G with C. A cannot pair with G, C or A and so on. This complementarity means that the sequence of one strand defines the sequence of the other: a G in one strand means that the other strand will always have a C in that position, etc. The consequence of this, from the cell's perspective, is that it will always have two copies of the information encoded by the DNA sequence. This is important because if one strand is damaged, the other can direct the repair and no information is lost.

The complementarity arises through hydrogen bond interactions between the bases. Looking at the structure of the paired bases reveals that the positions of these hydrogen bonds occur at the same relative positions between the functional groups of the different bases (Fig. 1.1). However, the pattern of hydrogen bond donors (the group with the H-atom) and acceptors (the interacting partner with no H) is different, so that if the bases are swapped around, H-bonds cannot be made (H-bonds cannot form between 2 donors or 2 acceptors). There are 2 H-bonds between A and T, and 3 H-bonds between G and C. This means that GC base pairs are more stable and thus harder to separate.

As two complementary strands of DNA come together the base pairs will interact with each other to form hydrogen bonds. The planar structure of the aromatic bases means that they will 'stack' on top of each other, each phosphate on the backbone has a negative charge and they are thus repelled from each other, while the sugar provides a linking group of a fixed length. The attraction of the stacking bases in each strand is balanced by the repulsion of the phosphates within the limit of the deoxyribose link. The net result is that each base is offset by around 34° from the previous one and the two strands twist into a double helix (Fig. 1.1).

1.2.2 *DNA replication*

The ability to reproduce is one of the key features of life and DNA replication lies at the heart of this process. The complementarity of the two DNA strands means that it is possible for the strands to separate and for each one to act as a template in the synthesis of a new DNA strand, so that two new daughter DNA helices are produced. Each one of these daughter DNA molecules contains one of the strands of the parental DNA double helix.

Cytosine (C)
Guanine (G)
5' Phosphate
3' Hydroxyl
Thymine (T)
Adenine (A)
Phosphate
Base
Sugar
3' Hydroxyl
5' Phosphate

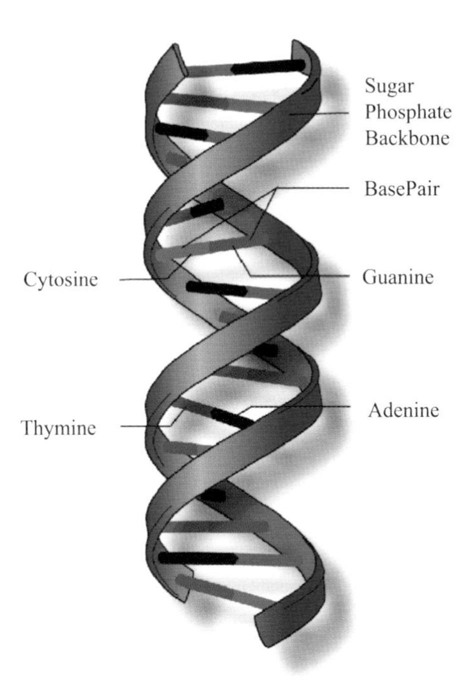

Fig. 1.1 The chemical structure of the nucleotide base pairs and the double-helical structure of DNA. (Courtesy: National Human Genome Research Institute)

DNA is synthesised by enzymes called DNA polymerases from component deoxyribonucleotide triphosphates (dNTPs). The triphosphate is a chemical group that effectively acts as an energy source for the reaction. This is a common feature in nature, and a different nucleotide — ATP — functions as a near universal energy source within living systems. DNA polymerases are able to select the correct nucleotide by its ability to base pair with the template strand. It then catalyses its incorporation into the growing strand, releasing a diphosphate (Fig. 1.2).

Another key feature of DNA structure is that the strands have a polarity as the chemistry of the backbone means that they are not the same at each end; they are denoted as the 5′ and 3′ ends. Double-stranded DNA always forms an anti-parallel configuration. This has very important implications for DNA replication, since DNA polymerases can only add on to the 3′ end. Furthermore, DNA polymerases can only extend an existing chain; they cannot initiate synthesis on single-stranded DNA. In a cell, replication is initiated by an enzyme called primase, which synthesises a short RNA primer *in situ* so that DNA polymerase can begin the process. This also enables a greater degree of control on DNA replication. *In vitro*, we also use short sequences to initiate DNA synthesis, but we use synthetic DNA oligonucleotides. It turns out that this is a very useful feature of DNA polymerases as it means we can direct the synthesis of DNA very precisely.

1.2.3 *PCR*

The polymerase chain reaction (PCR) is one of the most useful techniques in molecular biology and it is worth studying since understanding it requires understanding all of the key features of DNA structure and replication. PCR is a method for the exponential amplification of DNA. Its value lies in its speed and specificity; one can readily make large quantities of a specific DNA fragment from vanishingly small quantities of the original template sequence. It relies on DNA polymerases that have been isolated from bacteria that live in hot springs or ocean sea vents; these extremophile organisms can live in temperatures in excess of 100°C. As a consequence their DNA polymerases are also stable at high temperatures, a key feature that is required due to the thermal cycling of the reaction.

The specificity comes from the oligonucleotides that are used to prime the DNA polymerase. Two primers are required, one for each end of the DNA. The reaction operates through temperature cycling: the first stage requires heat denaturation of the DNA; as it is then cooled the primers will anneal to their complementary sequence on the DNA template. DNA synthesis is then directed from the 3′ end of the primers and the reaction is heated to 72°C, the optimal temperature of the thermostable DNA polymerases. The reaction is then heated to 95°C to melt the newly synthesised double-stranded DNA back to single strands, allowing for another round of priming and synthesis. The doubling nature of the reaction means that the specific DNA fragments increase in an exponential order so after 25–30 cycles a very large quantity of DNA has been synthesised (Fig. 1.3 and Appendix 1).

The general applicability of PCR has increased as new DNA polymerases have been isolated and engineered. The first thermostable DNA polymerase, *Taq*, had relatively

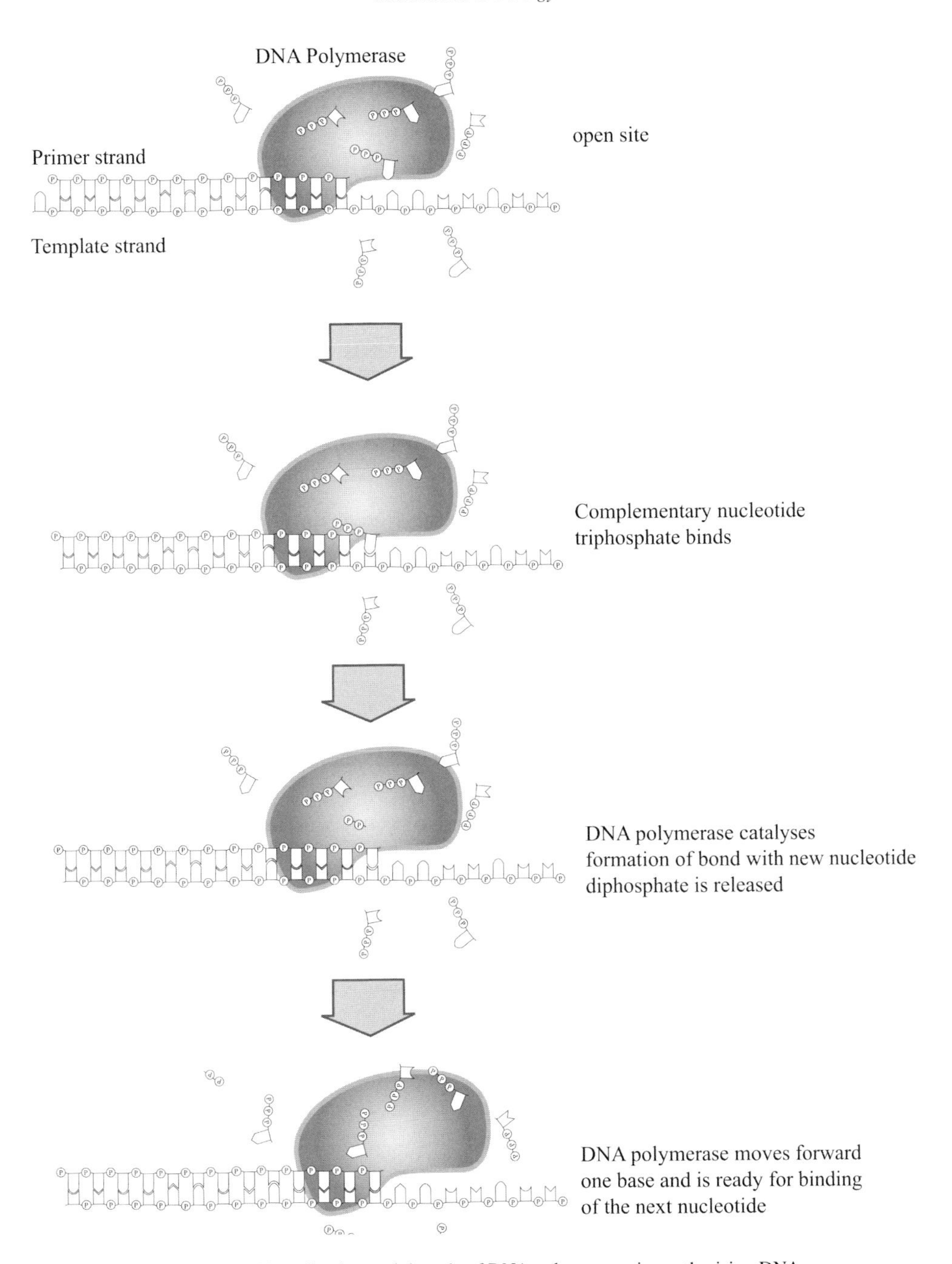

Fig. 1.2 DNA replication and the role of DNA polymerases in synthesising DNA.

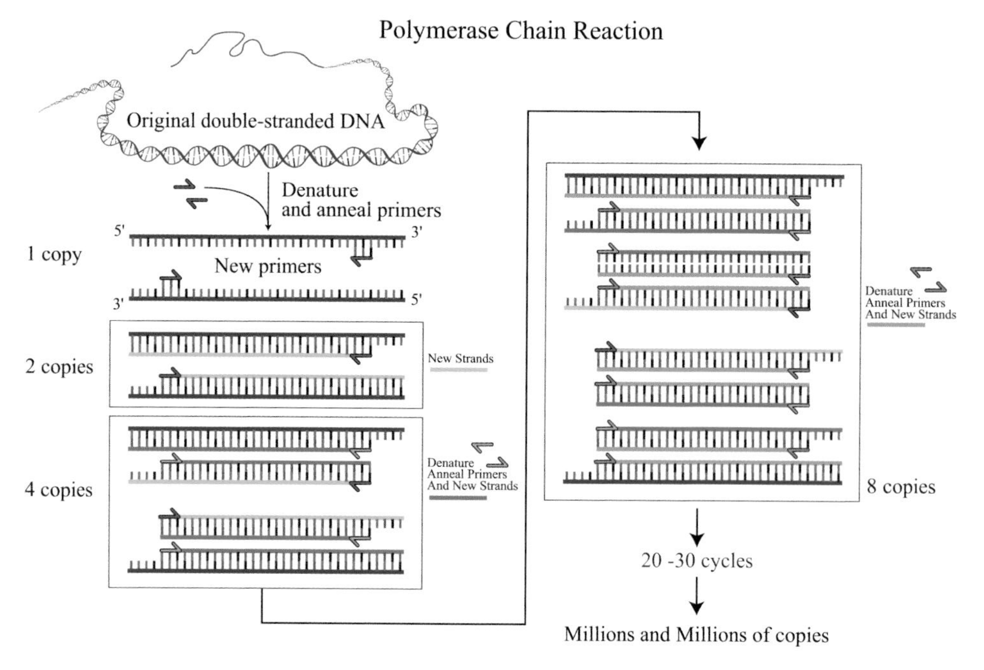

Fig. 1.3 The polymerase chain reaction. (Courtesy: National Human Genome Research Institute)

low fidelity, so amplification of sequences frequently led to mutations in the products. New DNA polymerases have been isolated from other sources that contain proofreading domains (a 3′–5′ exonuclease activity that is able to remove mismatched bases), which has significantly increased the accuracy with which DNA can be amplified. Other polymerases such as Phusion™
indexPhusion™} have been engineered to contain a DNA binding domain that increases the processivity of the polymerase, increasing the length of DNA that can readily be made by PCR. This increased accuracy means that PCR can now be adapted to many more cloning techniques that require amplification of the vector backbone (Ellis, Adie and Baldwin, 2011).

1.3 Information Flow in Biology

DNA is the information store, but the majority of functions within the cell are performed by proteins. The question thus arises of how information leads to function. The flow of information within biological systems forms what is known as the 'central dogma' (Fig. 1.4). Information is stored in the form of DNA. When required, the message within DNA is transcribed into an intermediate messenger molecule, RNA, before being translated into the final product, protein. The fundamental unit of hereditary information is known as a gene. Most genes are a stretch of DNA that codes for a protein, although some genes will produce functional RNA molecules rather than proteins. A gene should also be considered to include the regulatory elements required to control it. The basic structure of a gene is depicted in

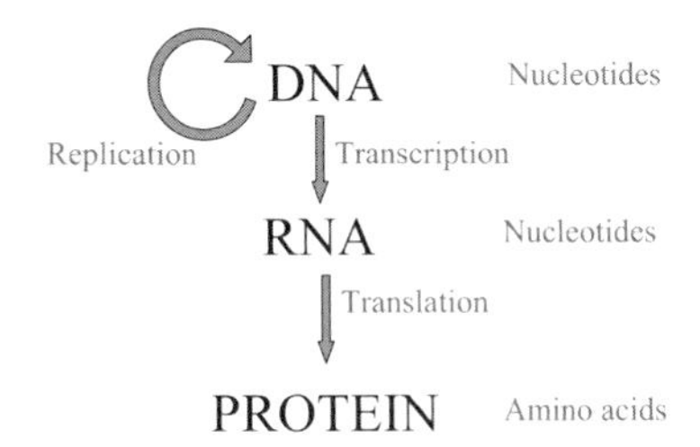

Fig. 1.4 The central dogma of molecular biology.

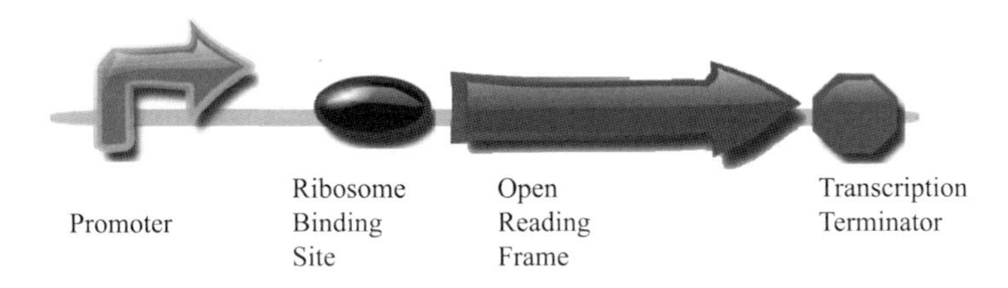

Fig. 1.5 The basic structure of a protein-coding gene.

Fig. 1.5. When a gene is turned on, the protein (or other functional molecule) is produced or expressed; when it is turned off it is not produced. More detail on these control mechanisms is provided in the next section.

The messenger molecule is composed of nucleotides similar to those found in DNA, called RNA (ribonucleic acid). The chemical structures of these two molecules are almost identical, the key difference being the 2'OH on the sugar moiety of RNA; the lack of this group in DNA is what gives the deoxy-designation. The presence of this extra OH means that RNA can form more complex structures; it is also less stable. In addition, RNA uses U (uracil) in place of T, the difference being the presence of a methyl group in thymine.

The conserved nature of the bases between RNA and DNA means that they can base pair in an identical way to DNA bases. This feature is used when transcribing the information in DNA into the messenger RNA (mRNA) molecule. During transcription a stretch of DNA known as a gene is transcribed into mRNA. That is, a mRNA molecule is synthesised *in situ* based on complementary base pairing between the RNA and DNA nucleotides. The key feature of a gene is that the sequence of DNA bases code for a protein. The complementary base pairing between RNA and DNA means that this code is retained in mRNA.

Once mRNA has been synthesised it fulfills its role as a messenger by taking the 'message' to the ribosome. This is a very large macromolecular complex composed of proteins and RNA molecules and it is here that the message is translated from nucleotides into amino acids, the building blocks of proteins. The end result is a protein where the amino acids have been put together in a specific sequence designated by the DNA sequence. Thus DNA codes for proteins.

RNA codon table

1st position	2nd position: U	C	A	G	3rd position
U	Phe	Ser	Tyr	Cys	U
	Phe	Ser	Tyr	Cys	C
	Leu	Ser	stop	stop	A
	Leu	Ser	stop	Trp	G
C	Leu	Pro	His	Arg	U
	Leu	Pro	His	Arg	C
	Leu	Pro	Gln	Arg	A
	Leu	Pro	Gln	Arg	G
A	Ile	Thr	Asn	Ser	U
	Ile	Thr	Asn	Ser	C
	Ile	Thr	Lys	Arg	A
	Met	Thr	Lys	Arg	G
G	Val	Ala	Asp	Gly	U
	Val	Ala	Asp	Gly	C
	Val	Ala	Glu	Gly	A
	Val	Ala	Glu	Gly	G
	Amino Acids				

Ala: Alanine
Arg: Arginine
Asn: Asparagine
Asp:Aspartic acid
Cys:Cysteine
Gln: Glutamine
Glu: Glutamic acid
Gly: Glycine
His: Histidine
Ile: Isoleucine
Leu: Leucine
Lys: Lysine
Met: Methionine
Phe: Phenylalanine
Pro: Proline
Ser: Serine
Thr: Threonine
Trp: Tryptophan
Tyr: Tyrosisne
Val: Valine

Fig. 1.6 The genetic code table. Amino acids are coded by three nucleotides (codons). The information is carried to the ribosome, which synthesises the protein encoded by the messenger RNA (mRNA). The codons are therefore usually shown in RNA nucleotides, where Uracil (U) is used in place of Thymine (T). (Courtesy: National Human Genome Research Institute)

1.3.1 *The genetic code*

The sequence of bases A, G, C and T provides the basis for information storage within DNA. This DNA sequence codes for protein sequence, but proteins are composed of 20 different amino acids, therefore the coding from nucleotides to amino acids is not one-to-one. It is, in fact, a triplet code, so three nucleotides code for a single amino acid (Fig. 1.6). The triplet code means that DNA sequence can be read in three different 'frames' on each strand (six in total; Fig. 1.7).

The triplet code means that there is redundancy, and all but two amino acids are coded for by multiple three nucleotide sequences (codons; Fig. 1.6). The exceptions are tryptophan and methionine, which have unique codons and this reflects their reduced frequency of occurrence in proteins.

Methionine is also the initiating amino acid in all bacterial proteins. The start signal for protein synthesis comprises the position where the ribosome binds the mRNA (ribosome binding site, RBS or Shine-Dalgarno sequence) and the first ATG codon, which must be close to the RBS. Reading the first codon establishes the reading frame of the mRNA, so having a unique codon for the initiating amino acid reduces ambiguity.

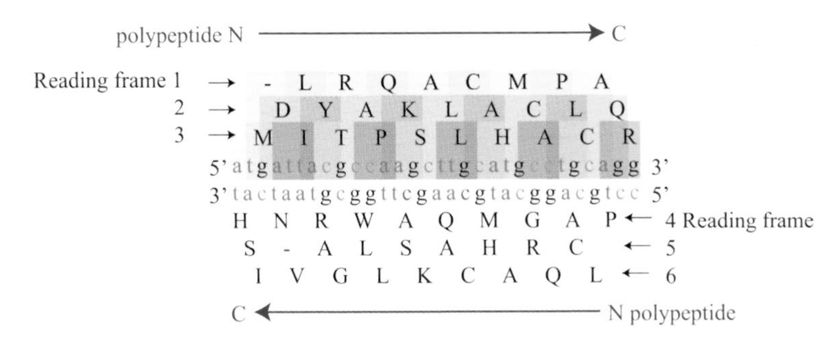

Fig. 1.7 A short section of double-stranded DNA. There are six possible coding frames on double-stranded DNA. The top strand codes for a protein. The beginning of the protein sequence is the ATG Methionine codon in reading frame 3. By shifting the reading frame by 1 bp steps, it can be seen that there are three other possible reading frames for the same piece of DNA. In reality these do not code for a protein because they do not satisfy the requirements of an open reading frame (they are not preceded by a promoter, are not uninterrupted by stop codons and do not finish with a stop codon). The other DNA strand also has three other possible reading frames. As the polarity of the DNA is reversed on the partner strand, these read in the opposite direction.

The degeneracy in the genetic code also leads to differences between different organisms. Each organism has evolved within its own niche and its translation machinery has typically developed a narrower set of codons that are regularly used, or are used under particular growth conditions. Since this codon usage varies from organism to organism, a gene that originates from one organism and is transcribed in another may be poorly translated as the codons of the gene are not regularly used in the new host.

The process by which the nucleotide sequence of the mRNA molecule is translated into protein (shown in Fig. 1.8) again requires intermediate molecules, this time to deconvolute the nucleotide sequence into amino acid sequence. These intermediaries are called transfer RNA molecules (tRNA) because they transfer amino acids onto the growing protein. As the name suggests, these molecules are composed of RNA, but are linked to amino acids. Cells contain different types of tRNA molecules, and each type can only be linked with a single amino acid. This highly specific reaction is catalysed by tRNA synthetase enzymes.

Because tRNAs are composed of nucleotides, they are able to base pair with the nucleotides of the mRNA, in the same way that DNA bases pair. The tRNAs are structured in such a way that a triplet of RNA bases is presented to the mRNA template, thus reading the genetic code and presenting the correct amino acid to the ribosome for incorporation.

1.3.2 *Proteins*

The great majority of functional and structural roles within organisms are carried out by proteins. They have a truly remarkable repertoire of properties and functions. They can form structural materials such as spider silk and keratin, which makes up hair and nails; they can act as motors and transduce chemical energy into mechanical energy, like myosin in muscles; there are many enzymes which catalyse highly specific chemical reactions and there are a large number of regulatory proteins. Regulation can occur at many levels from

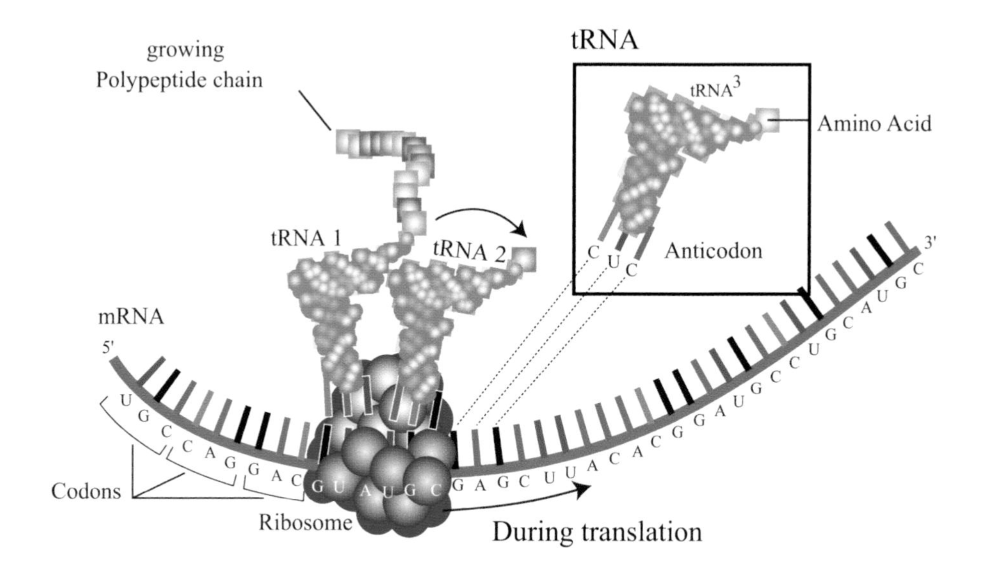

Common ways of illustrating tRNA

3'

5'

D loop

T loop

Anticodon loop

Anticodon

Anticodon

Fig. 1.8 Translation of mRNA into protein. The codons are read by transfer RNA molecules (tRNA). The anticodon on the tRNA is able to base pair with the mRNA codon. The amino acid is attached at the $3'$ end of the tRNA and when bound to the codon it brings it into the correct orientation for peptide bond formation, elongating the protein by one amino acid. (Courtesy: National Human Genome Research Institute)

hormones such as insulin, which regulates blood sugar levels, down to transcription factors which, as we will see later, control genes.

The diversity of proteins arises from a very limited set of building blocks: the 20 amino acids. The name amino acid comes from the fact that these molecules possess an amine (NH_2) and a carboxylic acid (COOH). These two functional groups are able to join chemically to

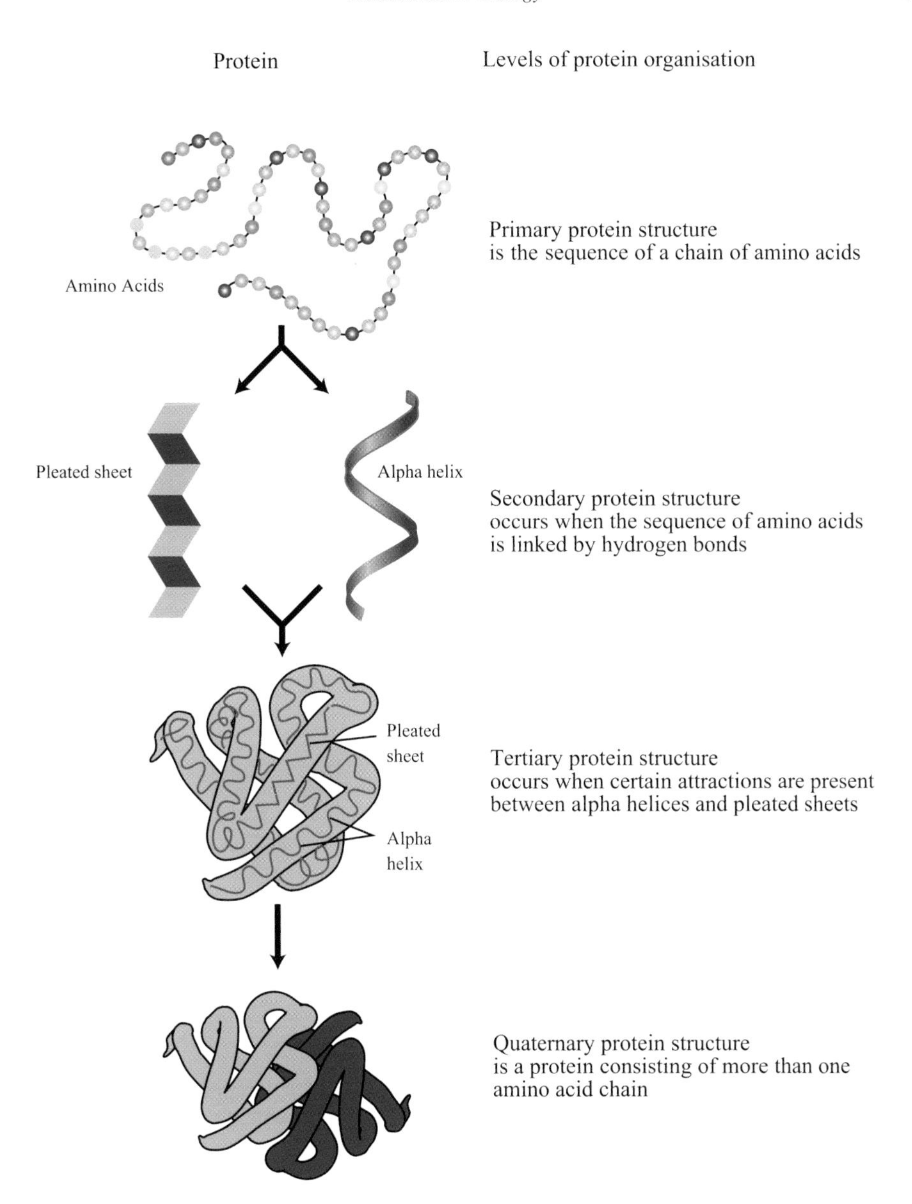

Protein
Levels of protein organisation
Amino Acids
Primary protein structure
is the sequence of a chain of amino acids
Pleated sheet
Alpha helix
Secondary protein structure
occurs when the sequence of amino acids
is linked by hydrogen bonds
Pleated
sheet
Alpha
helix
Tertiary protein structure
occurs when certain attractions are present
between alpha helices and pleated sheets
Quaternary protein structure
is a protein consisting of more than one
amino acid chain

Fig. 1.9 Amino acids and the primary to quaternary levels of protein structure.

form a peptide bond, with the elimination of a water molecule (H_2O). This reaction is catalysed in the ribosome and leads to the formation of a linear string of polymerised amino acids (polypeptide, or protein). As we have seen, the precise sequence of the amino acids is dictated by the DNA sequence. The functional properties of proteins do not arise directly from the linear sequence of amino acids (also called the primary structure), but rather the three-dimensional form that they adopt (Fig. 1.9). The folding of proteins into a three-dimensional form is an energetically favourable process, but the intricacies of how this is governed are highly complex and thus very difficult to predict. It is still not possible to predict computationally the three-dimensional structure of a given protein sequence. The difficulty in understanding the protein folding problem comes back to the incredible diversity of proteins. An average-sized protein would have around 300 amino acids with only 20 different amino acids, which means there are 20^{300} possible sequences. To put this into perspective, this is a substantially larger number than the number of atoms in the visible universe. This is of course the basis of the huge diversity of protein function, but it also highlights the difficulty of engineering proteins for new functions.

Despite this complexity, huge advances have been made in understanding the structural basis of proteins and their functional properties. These insights have been primarily due to X-ray crystallography, which has enabled atomic resolution structures of proteins to be determined. There are now over 60,000 protein X-ray structures that have been solved, and other techniques such as nuclear magnetic resonance (NMR) and electron microscopy are also making significant contributions to our understanding of protein structure and function.

The detailed atomic resolution of crystal structures has greatly facilitated our ability to rationally engineer proteins, since we can see the structural basis of specific interactions. However, this is still largely a 'hit-and-miss' affair, reflecting the complexity of atomic interactions that make a functional protein or enzyme. The great depth of understanding of protein structure has also led to advances in computational techniques to predict protein structure, perhaps best exemplified by the CASP competition (Critical Assessment of techniques for Protein Structure Prediction).

1.4 Controlling the Flow of Information in Biology

One of the surprises that became evident with the sequencing and annotation of genomes was how few genes are present in higher organisms, which typically contain around 25,000 open reading frames. By comparison, the well-studied *E. coli* prokaryote contains around 4,800. The difference in the number of open reading frames does not seem to reflect the increase in complexity between humans and *E. coli*; however, what differs substantially is the number of genes that are responsible for control functions, which are proportionally much higher in eukaryotes (Mattick, 2004). This underlies the importance of control functions in complexity.

The principle flow of information in biological systems determines how information is utilised at required times and in response to external signals. This essentially governs when and where a gene is expressed. In eukaryotes this is responsible for the process by which cells specialise and become skin cells or brain cells, for example. In prokaryotes it enables the bacterium to respond to external signals and change its gene expression pattern in response to new food sources, for example. The aim of synthetic biology is to predictively design and build new biological systems that, for example, programme a phenotypic response or produce a chemical compound. It has become increasingly evident that the fine control of the components of any biological pathway is critical for a successful outcome (Pfleger *et al.*, 2006; Pitera *et al.*, 2007).

1.4.1 *Transcriptional control*

The primary step at which control takes place is the transcription of DNA into mRNA. This process is performed by RNA polymerase and is a complex process involving several protein subunits. The complexity enables fine control of the process, but also makes understanding the precise mechanism rather difficult. Prokaryotes (bacteria) typically have much simpler systems than eukaryotes (higher organisms whose cells contain a nucleus), and consequently prokaryotic systems are much better understood. For this reason we will provide here a basic overview of prokaryotic transcription, although the reader should be aware that there are differences between prokaryotic and eukaryotic systems and should they require more detailed information then they are advised to consult more specialised texts.

A gene consists of the stretch of DNA that codes for a protein, known as the open reading frame (ORF), plus the regulatory elements that control its expression, that is when it is turned on and off. The most important control element is the promoter which can be considered as the genetic switch. In prokaryotes promoters can either control a single gene, or they can control multiple genes in what is called an operon. Operons have all the genes for a particular function clustered together so that their expression can be coordinated from a single promoter.

The principle enzyme responsible for transcription of a gene is RNA polymerase. The polarity of DNA is also very important in transcription because, as in replication, RNA polymerases only work in a $5'$ to $3'$ direction, and consequently genes are always read $5'$–$3'$ according to the 'coding strand'. The other DNA strand acts as the 'template strand' for RNA polymerase, which acts in an analogous manner to DNA polymerase, making a new polynucleotide chain that is complementary to the template. Hence the newly synthesised mRNA has the same sequence as the coding strand (Fig. 1.10).

RNA polymerase knows where to start transcribing a gene because it interacts with the promoter, which is a stretch of DNA upstream (on the $5'$ side) of the gene. RNA polymerase binds the promoter before it begins transcription and consequently the control of transcription is centred around controlling RNA polymerase activity. This involves a number of other proteins, the most fundamental of which are the sigma (σ) factors. Prokaryotic RNA polymerases do not work in the absence of a σ factor, and they provide a mechanism for global gene regulation that works under different environmental conditions: *E. coli* has seven different σ factors (Table 1.1). The complex of all the RNA polymerase subunits, including the σ factor, is known as the holoenzyme.

Sigma factors are intimately involved with the binding of RNA polymerase to a promoter. Promoters are DNA sequences upstream of where transcription is initiated (denoted as the $+1$ position). There are certain regions where it can be seen that the DNA sequence is conserved between different genes. This conservation of sequence underlies the importance this sequence has in transcription; if it were mutated then the promoter would not work as well, hence the sequence is conserved during evolution. There are two critical regions where prokaryotic RNA polymerase holoenzyme binds the DNA and recognises the promoter. These regions are 10 and 35 bases upstream of the transcription initiation site, and so are known as the -10 and -35 sites (Fig. 1.11).

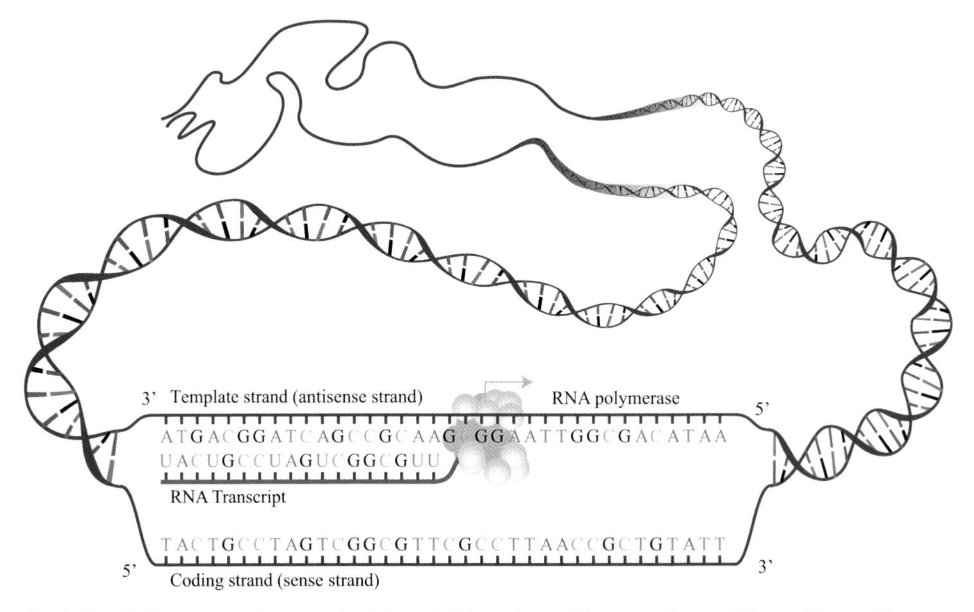

Fig. 1.10 Coding and template strands during mRNA synthesis. (Courtesy: National Human Genome Research Institute)

Table 1.1 *E. coli* σ factors.

Factor	Function
σ70 (RpoD)	The 'housekeeping', or primary sigma factor, transcribes most genes in growing cells. Makes the proteins necessary to keep the cell alive
σ54 (RpoN)	The nitrogen-limitation sigma factor
σ38 (RpoS)	The starvation/stationary phase sigma factor
σ32 (RpoH)	The heat shock sigma factor is turned on when exposed to heat
σ28 (RpoF)	The flagellar sigma factor
σ24 (RpoE)	The extracytoplasmic/extreme heat stress sigma factor
σ19 (FecI)	The ferric citrate sigma factor regulates the fec gene for iron transport

The σ factor is largely responsible for the direct interactions with the -10 and -35 sites, and it can be seen that different promoters have different consensus sequences at these positions. This reflects interactions with different σ factors, so that the expression of different sets of genes can be differentially controlled through different σ factors. The sigma factors themselves are controlled by environmental conditions, and there are also anti-σ factors that can inhibit their action, and even anti-anti-σ factors that can restore activity. σ factors typically operate on a number of promoters distributed throughout the genome and can thus be thought of as global regulators of gene expression.

Sigma factors are not the sole means by which transcription is regulated. There are two other types of proteins that control genes: activators and repressors. As their names suggest, these can increase and decrease gene transcription respectively. These transcription factors

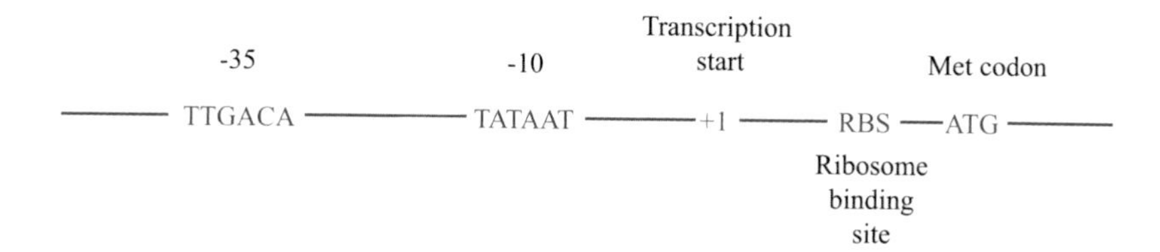

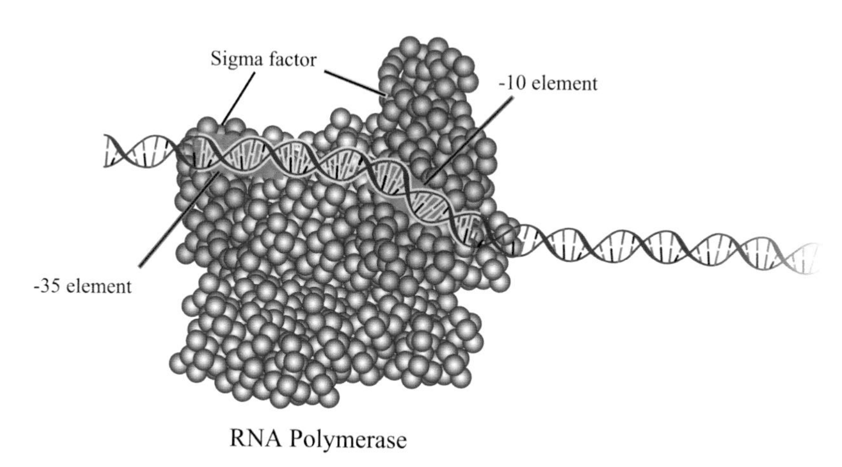

Fig. 1.11 Promoter structure and binding of RNA polymerase. Numbering is in relation to the position where transcription begins at the +1 site. The key elements of bacterial promoters are the −10 and −35 sequences, which interact with the σ-factors. Different sequences mean different promoters operate with different σ-factors (see Table 1.1). The RBS and first methionine codon of the open reading frame are downstream of the transcription start site.

are critically important in turning genes on and off, and typically work on a single operon or a limited number of promoters. Transcription factors thus act as discrete regulators of gene expression, in contrast to the global regulator σ factors.

It is often desirable for the control of transcription to have the characteristics of a switch: it is either on or off. It is worthwhile considering how this effect is achieved by the binding of proteins to a DNA sequence. Transcription factors typically bind specific DNA sequences, but the characteristics of a single protein binding a single site will give a hyperbolic dependency in binding. The effect of this is that the switch from on to off will occur relatively gradually as the concentration of a repressor increases; the switch will not be very responsive, and it will be hard to turn off completely. It is frequently observed that multiple proteins bind at more than one DNA binding site on the operator. When multiple proteins bind this can give rise to protein–protein interactions as well as protein–DNA interactions. This leads to the effect known as cooperativity: it is more favourable for the second protein to bind than the first because of the additional protein–protein contacts. As transcription factor concentration increases, this will lead to a sigmoidal (S-shaped) response in binding and the switch becomes much sharper. For those who wish to understand these important effects more, an insightful text on bacteriophage λ by Mark Ptashne is worthy of further study (Ptashne, 1992). A lot can also be learnt from modelling these interactions.

1.4.2 *Translational control*

Transcription produces messenger RNA (mRNA), which acts as the intermediate between the DNA 'archive' and the functional protein product. To produce the functional protein, the mRNA has to be translated from nucleotides into amino acids, which then fold to create a functional protein. This process is carried out by the ribosome, and the process is called translation.

In prokaryotes, the 5′ end of the mRNA molecule has a sequence that is recognised by the ribosome, called the ribosome binding site (RBS). The presence of this site allows the ribosome to bind the mRNA and it also positions it appropriately to begin translation at the first ATG (methionine) codon. The ribosome will then move down the mRNA, translating the codons into amino acids, and synthesising the polypeptide chain to create the protein according to the sequence encoded by the mRNA.

The RBS has a critical role in translation, and it is the most direct means of modulating how much protein is made from a single RNA molecule. mRNA molecules are relatively unstable and will have a half-life of around 2–3 minutes within a cell, so only a finite amount of protein can be made from a single mRNA molecule. The strength of the RBS will determine how much that is. A strong RBS has a tight interaction with the ribosome, so it will bind and begin translation effectively. Conversely, a weak RBS has a much lower affinity for the ribosome, which means it has a greater tendency to dissociate from the RBS and reduces the potential for protein synthesis, so less protein is made from that mRNA.

The RBS has a consensus sequence, which means that this is the representative sequence based on comparisons of known RBS sites. Slight variations in the sequence will still be functional, but will generally have higher or lower expression levels. In synthetic biology there may be a requirement to fine-tune protein levels, and the RBS provides a site of intervention that can be used to modulate the level of protein expression by slightly changing its sequence (Pfleger *et al.*, 2006).

One of the goals in synthetic biology is to have predictive design of biological systems. Understanding the level of translation for a particular protein can be an important parameter in predictive design. This presents one of the challenges faced when engineering biological systems, because the behaviour of a particular RBS will be dependent on the ORF that it is used with. This context dependence arises because the mRNA is a single-stranded molecule which can therefore fold and form secondary structures if there are homologous sequences in neighbouring regions. If an RBS site folds into a secondary structure then it will be unavailable for binding to a ribosome, which will only bind an open single-stranded RNA molecule, so protein expression will decrease.

One method for determining and rationalising these effects is to take a numerical approach. The RBS calculator determines the free energy difference between a folded mRNA transcript and the same mRNA bound to the ribosome (Salis, Mirsky and Voigt, 2009).

A number of parameters are considered in the calculation and it predicts the effective strength of a synthetic RBS site within a specified sequence context. This has demonstrated a correlation between the predicted and observed behaviour of a given DNA sequence to within a factor of 2–3. This is an important advance in considering how we tackle these types of problems, although it is also clear that the accuracy of the current predictive model is not sufficient for a rational design-based approach.

1.4.3 *RNA regulation*

It is becoming increasingly recognised that RNA is not merely a passive messenger molecule within the cellular milieu. RNA has a number of important catalytic and regulatory functions and it is thus integral to the overall working of the cell. Its vital role in many of the most fundamental cellular processes underlies its ancient heritage, and it is thought that RNA was the first genomic material and that an 'RNA world' preceded the evolution of the current DNA world (Atkins *et al.*, 2011). The evolution of DNA provided a more stable genetic material that facilitated the evolution of higher complexity organisms. The RNA legacy can still be observed in many vital cellular functions. For instance, it forms many of the components within the ribosome, and tRNA molecules are also integral to the process of protein synthesis.

RNA can also act as a catalyst (ribozymes) and many small RNA molecules are now thought to act as regulators of gene expression (Mattick, 2004; Atkins *et al.*, 2011). Within eukaryotes the discovery of siRNA (small interfering RNA) as regulators of gene expression has led to the development of valuable tools for manipulating gene expression as well as the discovery of new pathways for RNA processing and native gene regulation. It is also thought that prokaryotes use small RNA molecules to regulate gene expression, although they do not contain the same RNA processing machinery as eukaryotes.

It has been elegantly demonstrated that small RNA molecules can be engineered to interact with mRNAs and modulate the translation step (Bayer and Smolke, 2005). The short half-life of RNA molecules within prokaryotes means that specific degradation pathways are not required to specifically remove the targeted mRNAs. The short half-life of RNA also makes them an attractive target for regulatory processes, since the response can be rapidly turned on and off (Win and Smolke, 2007). Using proteins is more costly in terms of cellular resources; they are relatively slow to make and can have exceedingly long lifetimes, leading to slow response times.

An interesting variant of RNA control elements are riboswitches (Tucker and Breaker, 2005). These are sequences in the 5′ end of an mRNA that modulate the translation through binding of small metabolite molecules. When the mRNA binds to the metabolite, specific structural complexes are made that sequester the RBS and prevent translation. Frequently riboswitches are control elements that regulate the enzymes responsible for the metabolic pathways, providing feedback mechanisms of control. Recently riboswitches have been engineered by synthetic biologists to provide programmable control mechanisms (Callura *et al.*, 2010).

Reading

Atkins, JF, Gesteland, RF and Cech, TR. (2011). *RNA Worlds: From Life's Origins to Diversity in Gene Regulation.* Cold Spring Harbor Press, New York.

Bayer, TS and Smolke, CD. (2005). Programmable, ligand-controlled riboregulators of eukaryotic gene expression. *Nat Biotechnol* 23: 337–343.

Callura, JM, Dwyer, DJ, Isaacs, FJ, *et al.* (2010). Tracking, tuning, and terminating microbial physiology using synthetic riboregulators. *Proc Natl Acad Sci U S A* 107: 15898–15903.

Ellis, T, Adie, T and Baldwin, GS. (2011). DNA assembly for synthetic biology: from parts to pathways and beyond. *Integrative Biology* 3: 109–118.

Mattick, JS. (2004). RNA regulation: a new genetics? *Nat Rev Genet* 5: 316–323.

Pfleger, BF, Pitera, DJ, Smolke, CD, *et al.* (2006). Combinatorial engineering of intergenic regions in operons tunes expression of multiple genes. *Nat Biotechnol* 24: 1027–1032.

Pitera, DJ, Paddon, CJ, Newman, JD, *et al.* (2007). Balancing a heterologous mevalonate pathway for improved isoprenoid production in *Escherichia coli*. *Metab Eng* 9: 193–207.

Ptashne, M. (1992). *A Genetic Switch.* Blackwell Scientific, Cambridge, MA.

Salis, HM, Mirsky, EA and Voigt, CA. (2009). Automated design of synthetic ribosome binding sites to control protein expression. *Nat Biotechnol* 27: 946–950.

Tucker, BJ and Breaker, RR. (2005). Riboswitches as versatile gene control elements. *Curr Opin Struct Biol* 15: 342–348.

Win, MN and Smolke, CD. (2007). RNA as a versatile and powerful platform for engineering genetic regulatory tools. *Biotechnol Genet Eng Rev* 24: 311–346.

CHAPTER 2

Basic Concepts in Engineering

2.1 Introduction

This chapter describes basic concepts in engineering which have direct application in synthetic biology. Systematic design is a key approach in synthetic biology. This is described, together with the concept of parts, devices and systems, the synthetic biology design cycle and modelling techniques. Other important areas relating to the application of engineering methods include the registry of parts and part characterisation. Underlying this is the concept of web-based information systems and data templates for parts data. Another important aspect of synthetic biology is the development of platform technology, which is discussed in relation to the information system and the concept of BioCAD. The final section of the chapter deals with some example applications.

Francis Collins, the director responsible for the Human Genome Project (see Chapter 3) within the National Institute of Health, stated in a lecture approximately two years after the publication in *Nature*, that 'the initial sequencing of the human genome would not have been possible without the extensive use of ICT and computers'. This statement encapsulates how the confluence of engineering, physical science, ICT and computing on the one hand and biology on the other has resulted in the development of a new discipline of synthetic biology.

2.2 Systematic Design

An important aspect of synthetic biology is the application of systematic design to the creation of biologically-based devices and systems. This approach is based upon the engineering principles of modularity, characterisation and standardisation. The concept of modularity is the ability to reduce a device or system to a number of component parts. In this approach each of the parts is characterised in detail. That is to say that the characteristics of the part and how it operates under different conditions are rigorously studied. Where possible, parts are standardised in order that, in principle, a series of standard parts can be combined to create a device. The power of this approach, which is used in many areas of engineering, is that a device does not need to be produced from scratch every time, but, rather, it can be created from existing standard parts (and as we will see, in some cases a combination of standard parts and new parts). These concepts, which are widely used in engineering and manufacturing, are currently being applied to synthetic biology. This is known as the Parts, Devices and Systems approach to synthetic biology design (see Chapter 5 for a biological explanation of parts in the synthetic biology context).

2.2.1 *The parts, devices and systems approach*

In current synthetic biology a hierarchy based on the Parts, Devices and Systems approach can be defined as follows:

(i) Parts (bioparts) — these encode biological functions (currently synthetically designed DNA).
(ii) Devices — these are made from a collection of parts (bioparts) and encode human defined functions (e.g. logic gates).
(iii) Systems — systems perform tasks, such as counting and, potentially in the future, intracellular control functions.

Hence, in principle, to construct a simple gene circuit comprising a promoter, ribosome binding site, protein coding sequence and terminator would consist of joining four sections of DNA (where each section is a part, i.e. a specific DNA sequence for the considered part).

2.2.2 *The synthetic biology design cycle and its role in systematic design*

A key aspect of synthetic biology is the application to biology of techniques which are normally used in engineering design and development. The essence of this approach is to define the specification of the part, device or system required and to develop a design which meets these specifications. This overall approach is part of what is known as the design cycle; this is illustrated in Fig. 2.1.

In Fig. 2.1, the specification step is followed by a detailed design step. One of the key differences between design today and that of the past is the ability to undertake detailed computer modelling. This is also true in synthetic biology. Comparisons are often made between the difficulty of designing biological systems and the design of electronic devices, such as transistors, 60 years ago. The difference today is the wide availability of large amounts of computer power, which makes it possible to carry out detailed computer modelling. This means that the expected behaviour of the part, device or system under development can be simulated in detail. The next stage in the cycle is implementation. In synthetic biology this normally means synthesising DNA and inserting it into a host cell (chassis). The next stage of the design cycle, testing and validation, is particularly important in synthetic biology because it is the host response to the insertion of new DNA which determines whether or not the specification and the design have been properly realised.

Another key aspect of the design cycle approach is that the development of a part, device or system can involve a number of iterations of the cycle, with each iteration refining the design and its implementation. Engineering systems, for example the A380 Airbus,

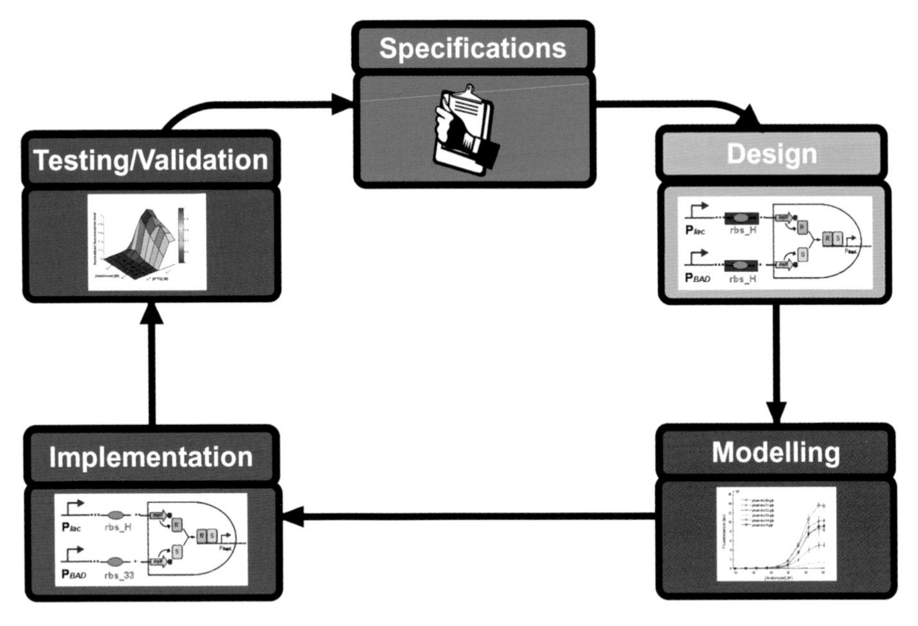

Fig. 2.1 The synthetic biology design cycle.

are based on standard devices, which are built from standard parts. In synthetic biology the field of electronics is sometimes used as a conceptual model. Taking the example of a simple audio amplifier, this would be designed using standard resistors, capacitors and transistors. The designer would have a set of specifications for the amplifier and look up manufacturers' handbooks (today, probably on the internet) to find component parts which meet the exact specifications which are required for the design. What is important here is to understand that a great deal of time and effort will have been devoted by component manufacturers to produce parts which exactly matched their specifications. In fact, it is common in engineering for a number of manufacturers to make and sell parts which exactly match the same specifications. It is important to note that the designer of the audio amplifier would not question for a moment that a commercially available part exactly matches its specification. Once built, tested and validated, the audio amplifier becomes a standard device built from standard parts — with its own specification sheet (the same approach applies to standard parts and devices in synthetic biology).

Taking this whole approach one step further, now suppose that the objective is to build a simple radio. Radios basically comprise three standard devices: a radio frequency, or RF stage; an intermediate frequency, or IF stage; and an audio stage (the audio amplifier). Hence, the aim would be to build the radio from these three standard devices. It may well be the case that each stage is manufactured and supplied by a different manufacturer but, because they are all standard, it is possible to connect the three stages together to form the radio. This is a very powerful approach and is routinely applied to the building of most engineering systems because it does not require everything to be designed and built from scratch.

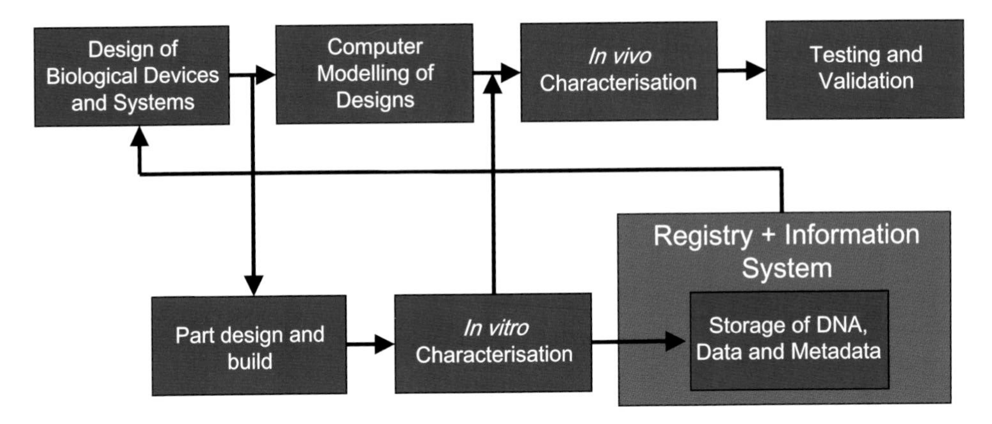

Fig. 2.2 Device design — incorporating the design cycle and the registry of parts.

A similar conceptual approach is used in synthetic biology. Here a number of bioparts are combined to produce a particular device (on the basis of a particular design). Figure 2.2 illustrates this process. The top line of the figure comprises a linear version of the synthetic biology design cycle. Referring to the first block (design of biological devices and systems) at this point, the bioparts which were used in the design are defined. On this basis, the registry is first interrogated to determine which standard bioparts are available. If additional bioparts are required then new parts need to be designed and built (as shown in the lower section of the diagram). Following the design and build the part is characterised and the characterisation data and metadata are added to the registry as a new part. The part is also then included in the fabrication of the new device.

2.3 The Registry and Part Characterisation

Earlier in this chapter the concept of the engineering approach to synthetic biology design was introduced in terms of parts, devices and systems. To summarise, the basic concept is one of using a combination of standard parts to produce standard devices, which are then combined to produce standard systems. This means that the standard parts must be thoroughly characterised so that their performance is carefully determined in relation to all the conditions under which the part will operate normally in the context of a particular type of chassis (cell). Hence, for example, the function of the part is defined on datasheets which comprise graphical, tabular and written information. In addition, metadata is usually provided. This is data about the context in which the data relating to the part has been obtained (experimental conditions, type of equipment etc). Metadata is often referred to as data about data.

Unlike the situation 20 years ago, datasheets in almost all fields today are stored electronically in databases. In the case of synthetic biology parts (bioparts), data and metadata are stored in a registry of parts. The best known registry is currently the so-called MIT Registry for the International Genetically Engineered Machine (iGEM) competition (see Chapter 8). This is a large registry containing many parts which have been designed for the iGEM competition. It does, however, have a number of problems. For example, the data tend to be more qualitative than quantitative; the parts do not always function as stated; and within the iGEM competition the focus has been on quantity rather than the quality of parts. In practice for this registry parts can be submitted by any user; they do not need to conform to a universal structure; they often contain missing data fields; and they are not automatically verified upon submission to the registry. There is, therefore, a widely held view that for serious synthetic biology design there is a need for a professional registry of parts.

2.4 Information Systems

Since the concept of a registry of parts is not one which is unfamiliar in a range of fields, it is possible to look at previous experience in other fields to try and design an ideal format for a professional registry of parts. For most registries of parts it is now the case that the part data and metadata are stored in an electronic database, which forms part of what is commonly known as a web-based information system. In synthetic biology, Imperial College has developed an information system called SynBIS (standing for Synthetic Biology Information System). The system is illustrated in Fig. 2.3. Referring to the figure, it can be seen that SynBIS is a web-based information system which comprises a fourlayer

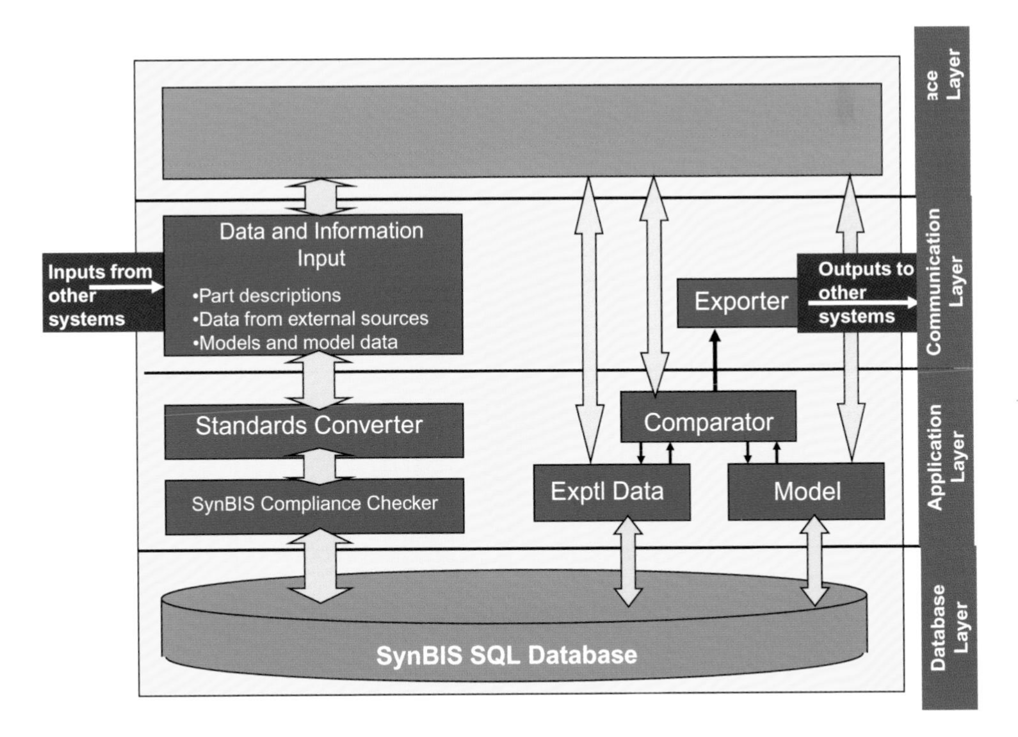

Fig. 2.3 The Synthetic Biology Information System (SynBIS).

architecture consisting of an interface layer (the web browser); a communication layer; an application layer (comprising specialist software); and a database layer (which comprises a structured query language (SQL) commercial database).

2.4.1 *The SynBIS system*

The specific implementation of the model can best be described in terms of the detailed diagram of SynBIS (Fig. 2.3). The left- and right-hand sides of the diagram are divided according to the following structure.

2.4.1.1 *Left-hand side of the system diagram (Fig. 2.3)*

This comprises the functionality to input various types of data and information into the system. How this is achieved will now be described in terms of each of the layers.

2.4.1.1.1 Interface (HTML) Layer

This comprises a web-browser (HTML) interface where the user can control the input of various types of data and information into the SynBIS database. These include the following:

(i) Part description (type, function, etc.).
(ii) Experimental data, from an external source — a typical example being characteristic data for a particular part.
(iii) Model description (i.e. the description of models in terms of their equations).
(iv) Data derived from specific models.

2.4.1.1.2 Communication Layer

The role of the Communication Layer on this side of the system diagram is to allow the information and data described in the previous section, above, to enter the system.

2.4.1.1.3 Application Layer

Once the data and information have been entered into the system at the Interface Layer, it is transferred to the Application Layer. Here there are two types of application software. The first converts known data types to the SynBIS data standard via standards converter software. The second set of software is conformance software, which checks that the data and information conform to the SynBIS template. If this is not the case then the system delivers a compliance report.

2.4.1.1.4 Database Layer

This comprises the SQL database, whose schema is the SynBIS template. All the information and data which is entered into the SynBIS system must conform to this template.

2.4.1.2 *Right-hand side of the system diagram (Fig. 2.3)*

2.4.1.2.1 Interface (HTML) Layer

As previously discussed, this comprises a web-browser (HTML) interface. However, in this section of the system diagram the user can control a number of system functions. These include:

(i) The display and manipulation of various types of data and information from the SynBIS database.
(ii) The export of data and information to other systems.
(iii) Running computer models and displaying the results.
(iv) Running comparisons of experimental data from the database with model results.

2.4.1.2.2 Communication Layer

The primary role of the Communication Layer on this side of the system diagram is to allow the information and data which result from comparing experimental data with model data to be exported to other systems.

2.4.1.2.3 Application Layer

On this side of the system diagram the Application Layer carries out a number of important functions:

(i) It enables the extraction, display and manipulation of data and information from the database.
(ii) It enables data and information (such as parameters) from the database to populate models for the purpose of simulation.
(iii) It controls the comparison of experimental data from the database with the results of simulations. Such comparative studies are important because their objective is to study how and why the experimental data differs from that predicted by the model simulation. On this basis new parameters can be fed to external systems for the purpose of optimising parts.

2.4.1.3 *Data and metadata*

As previously described, the information relating to a particular biopart is stored in the database under two separate headings. Hence, conceptually, it is important to understand the schema for the storage of such information in the database. Part type and chassis type are two key pieces of information associated with a particular biopart. Figure 2.4 illustrates the schema for the data and metadata for a particular part, in this example, a promoter. The basic information relating to the part is at the top of the schematic (part type and part number); below this are the characteristics of the part. In practice the characteristics may well comprise graphical information and, also, alphanumeric information. For a given part the scientists characterising the part will determine how many separate characteristics are

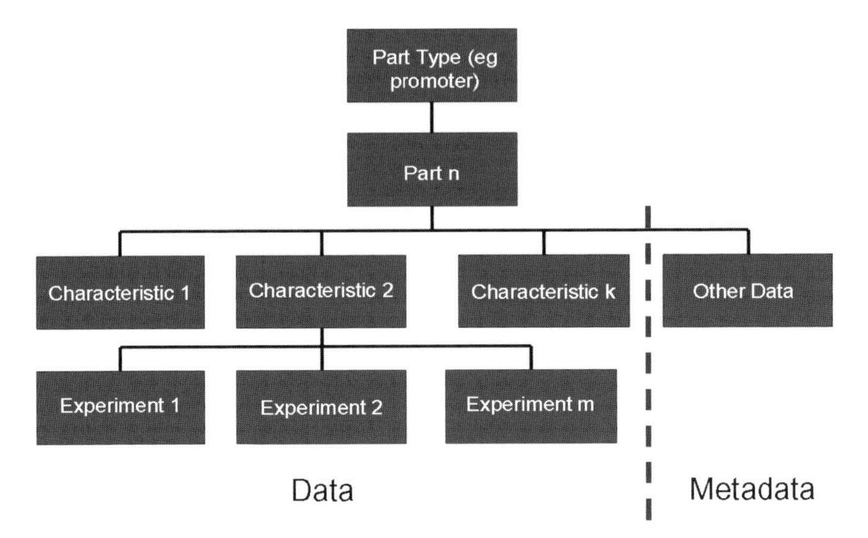

Fig. 2.4 A simplified version of the SynBIS Data Template.

needed to fully characterise the part. This will determine the nature of the experimental information and its extent.

To illustrate this further, Fig. 2.5 shows a part datasheet for the promoter LacUV5. Referring to the figure, it can be seen that in this particular case the datasheet comprises various types of graphical information — as well as images, sequences and diagrams. In addition, there is a significant amount of metadata. In relation to the metadata, the aim is to produce, as far as possible, a standard template for metadata for promoters (and similar metadata schema for other types of part). In reality, the metadata comprises a series of categories, each of which has associated active fields. For a particular case individual fields may or may not be populated.

2.5 The BioCAD Concept

Some of the key points which have been made so far can be summarised as follows:

(i) In synthetic biology design an important approach is to build standard systems from standard devices which comprise standard parts.
(ii) This is done in the context of the design cycle and systematic design approach.
(iii) Standard parts are stored in a registry of parts.

In the synthetic biology design process there is an interplay between the dry laboratory and the wet laboratory. It is very important to understand that normally, in practice, wet laboratory work takes far longer than work in the dry laboratory. Hence, modelling (dry lab work) is important because it can reduce the amount of time needed for part characterisation in the wet lab.

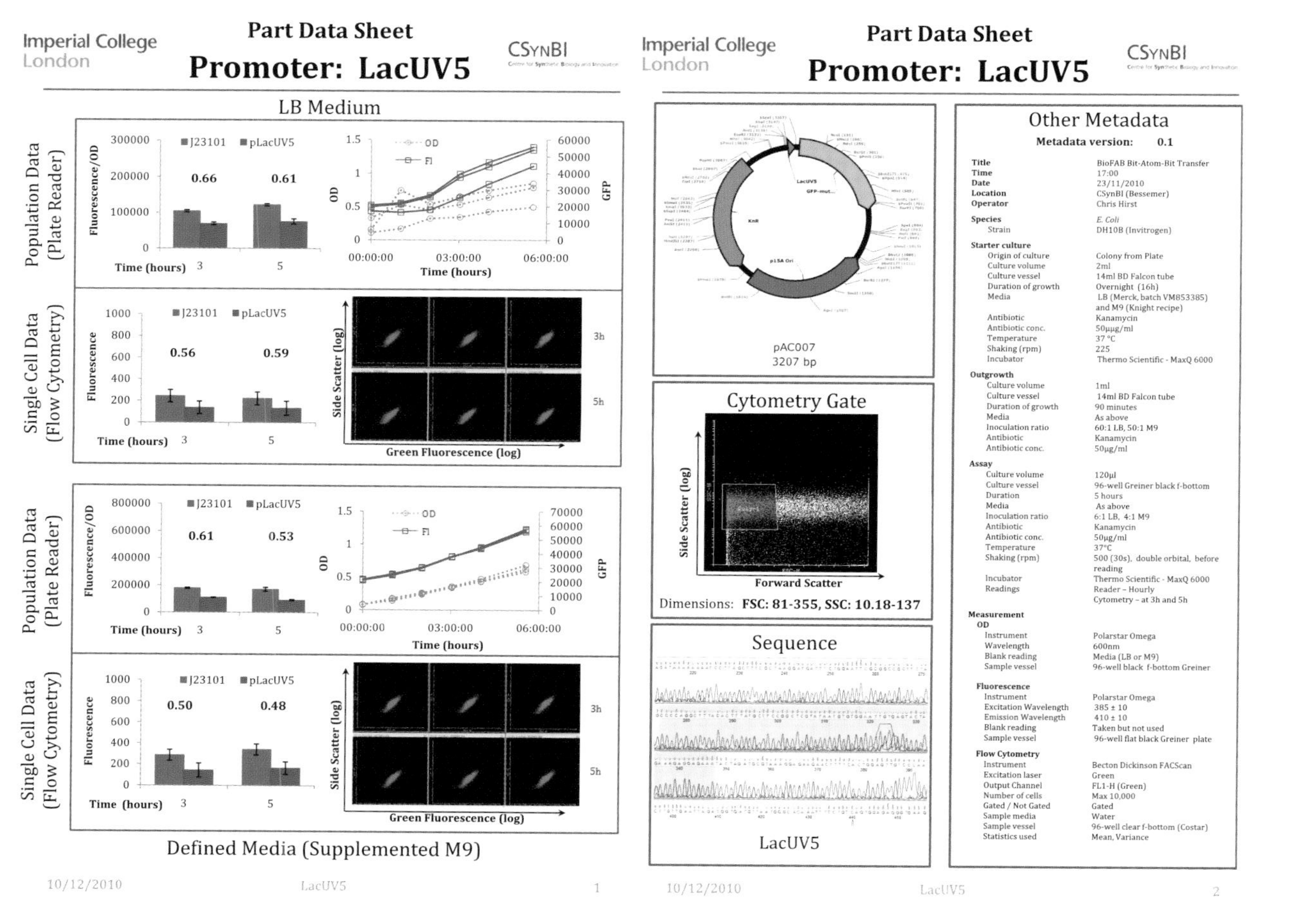

Fig. 2.5 An illustrative example of a promoter datasheet.

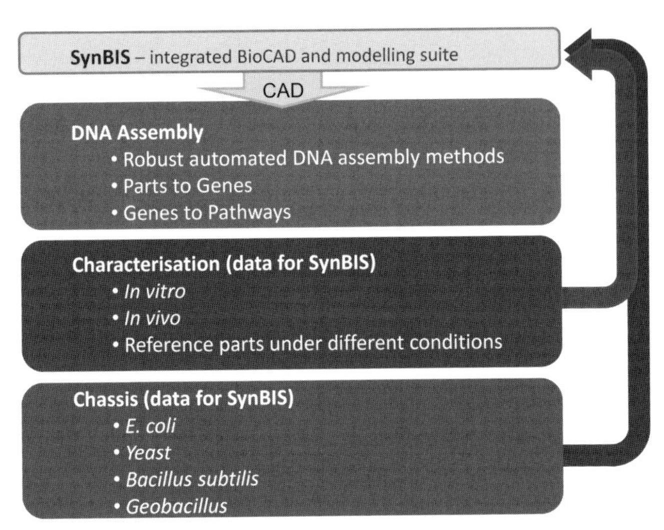

Fig. 2.6 Components of Platform Technology which are incorporated into SynBIS to provide an integrated BioCAD and modelling suite.

In this context (and again in relation to what is done in other fields), the concept of an information system which incorporates a Computer-Aided Design approach to synthetic biology, is being developed — this is often referred to as the BioCAD concept. Figure 2.6 illustrates the BioCAD concept in relation to its incorporation into SynBIS. BioCAD incorporates a number of aspects of synthetic biology design: DNA assembly, characterisation (data for SynBIS) and chassis (data for SynBIS). Hence, the concept is that a major amount of design and modelling work can be undertaken in the dry laboratory using BioCAD software.

More specifically, in relation to BioCAD, the systematic design approach in synthetic biology has many similarities to more general computer modelling. Referring to the left-hand side of Fig. 2.7, in computer modelling a block diagram of the design is obtained. Then for each block a mathematical description of the function of the block is derived. The entire block diagram and its associated mathematical descriptions are then converted to high level computer code using a high level computer language (such as C or C++). Within the computer the high level code is then converted to assembly code and then down to machine code (these last two steps are automatic).

On the right-hand side of Fig. 2.7, in synthetic biology a block diagram of the design is first obtained. The next step is to convert the basic block diagram into a biopart (or gene circuit) block diagram. At this point, computer simulations are run and the synthetic biology block diagram design and the associated bioparts block diagram are modified in relation to the results of the simulation. Once the biopart block diagram has been optimised, the next stage is to progress to assembly strategy software. This then leads to DNA assembly and ultimately to viability testing in the wet lab.

It is important to state that currently the BioCAD concept (as well as the associated BioCAD software) is still being developed. There are, however, a number of computer

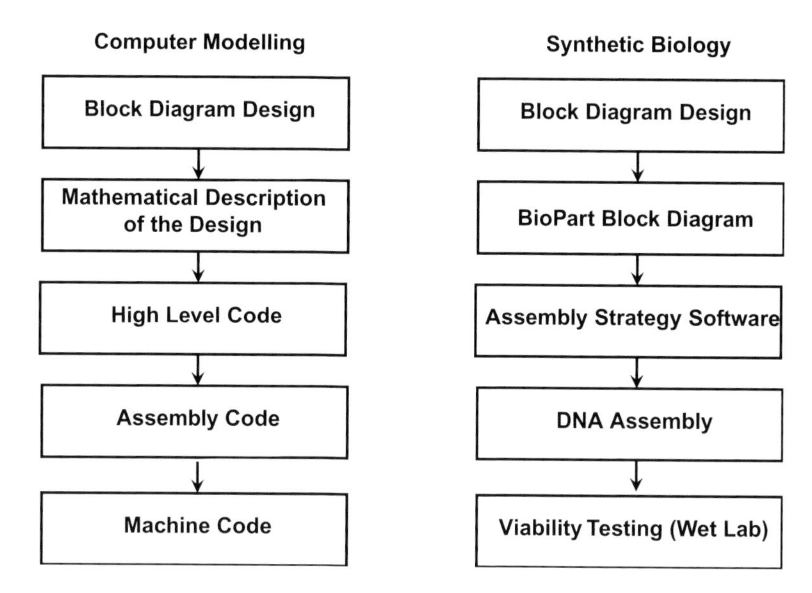

Fig. 2.7 A BioCAD Schema.

packages available which deal with various levels of the synthetic biology CAD block diagram. These include Clotho, GenoCAD, TinkerCell and a number of other open source packages.

2.6 Modelling

The device shown in Fig. 2.8 is a typical representation of a transcriptional regulatory device which comprises an inducible promoter, RBS, coding region and a terminator. Using the transcriptional regulatory device as an example, the following section describes the methodology of representing a biological device/system using block diagrams.

The device can be described by ordinary differential equations (ODEs). The same ODEs can also be used to describe devices such as inverters and switches. Referring to the equations, Eq. (1) describes the transcription of mRNA, which is activated in the presence of transcription factor $[W]$. In the example, $[W]$ is considered to be an inducer. Hence $\mu = +1$. A Hill kinetic function is used to model these dynamics. Consequently, Eq. (1) is nonlinear.

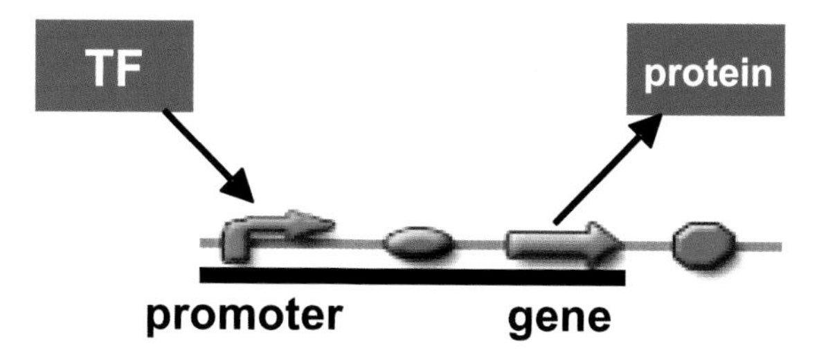

Fig. 2.8 A diagram of typical transcriptional regulatory device (TF refers to transcription factor).

This equation takes $[W]$ as the input and $[mRNA]$ as the output. Equation (2) describes the translation of protein $[P]$. The input to this process is the concentration of mRNA and the output is the concentration of P. This is treated as a linear equation.

$$\frac{d[mRNA]}{dt} = \frac{k_{tr}\cdot\left(\frac{W^n}{K^n}\right)^{\mu}}{1+\left(\frac{W^n}{K^n}\right)} - d_m\cdot[mRNA] \tag{1}$$

$$\frac{d[P]}{dt} = k_{tl}[mRNA] - d_p\cdot[P] \tag{2}$$

where

$[mRNA]$	mRNA Concentration
$[P]$	Protein Concentration
k_{tr}	kinetic constant of transcription
k_{tl}	kinetic constant of translation
$[W]$	inducer $\mu = +1$
	repressor $\mu = 0$
K	Activation/Repression Coefficient
n	Hill Coefficient
d_m	Degradation constant of mRNA
d_p	Degradation constant of Protein

The problem when analysing biological systems — or in the case of synthetic biology, designing biological devices and systems — is that as the system becomes more complex it becomes progressively more difficult to have an intuitive 'feel' for linking the mathematical description to the biology. This is a problem that has been faced on many occasions in the past. Fortunately, in engineering and physics there is a whole branch of mathematics called Systems and Control Theory, which has been developed over many years to address this problem. In fact the basis of the methods largely revolves around a mathematical theory which was pioneered by two French mathematicians: Pierre-Simon Laplace and Joseph Fourier at the beginning of the 19th century.

Their work is fundamental to much of modern engineering analysis. The essence of the Laplace approach is to convert or transform ODEs into what is called another domain — the Laplace Domain. In this domain, ODEs become algebraic equations, which are far easier to manipulate. Once a solution has been obtained in the Laplace Domain the final equation/solution is retransformed back into the Time Domain (i.e. the real world). The conversion of a function (e.g. an equation) from the Time to the Laplace Domain is done using the forward Laplace transform, *vis*:

$$F(s) = \int_0^{\infty} f(t)e^{-st}dt \tag{3}$$

where s is the so-called Laplace transform operator. Hence, the Laplace transform is used extensively to systematise the solution of ordinary differential equations.

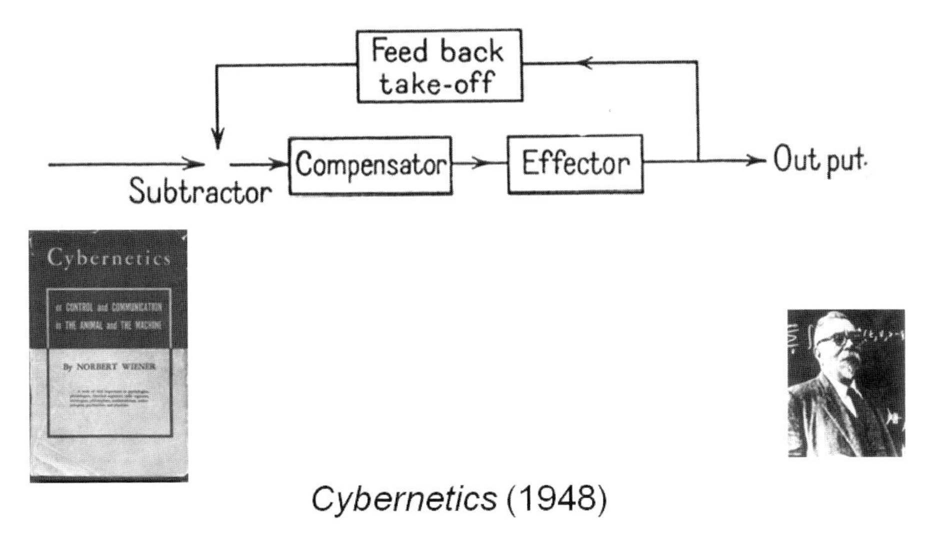

Fig. 2.9 Example of a feedback loop. (Courtesy: Weiner, N. (1961 [1948]). *Cybernetics*. MIT Press, Cambridge, MA).

2.7 Norbert Weiner

Sixty years ago two major building blocks were put in place in relation to systems and signal theory — the publication of Norbert Wiener's book, *Cybernetics*, in 1948, and the publication of Claude Shannon's work on information theory in the same year. In *Cybernetics*, Wiener established the mathematical basis for studying biological systems. Wiener's work, and that of others, has resulted in a major area of engineering science called systems theory. The main goal of systems theory is to define general principles that can be used to describe (and also design) systems regardless of their nature — biological, electrical, ecological, etc. This, coupled with signal theory (another very important area of engineering), has been widely applied in a range of fields, including biology in the form of systems biology. For example, systems theory is widely used in various industrial areas nowadays, such as the design and construction of aircraft control systems, information and telecommunication networks and economics.

2.8 Signal Theory

In addition to the use of systems theory, in order to describe biological systems there is often a need to study waveforms. Waveforms are often expressed in terms of the time evolution of a function; for example, the evolution of amplitude of a particular quantity against time. In many areas of science and engineering this is achieved by the application of signal theory. This area of theory can be used to analyse signals in biological systems (such as in cell signalling). In nature, signals associated with experiments (waveforms) are normally defined as being 'continuous'. Continuous in this context means that over the period during which the waveform is being recorded, it comprises finite values at all points in time (see Fig. 2.10, left side).

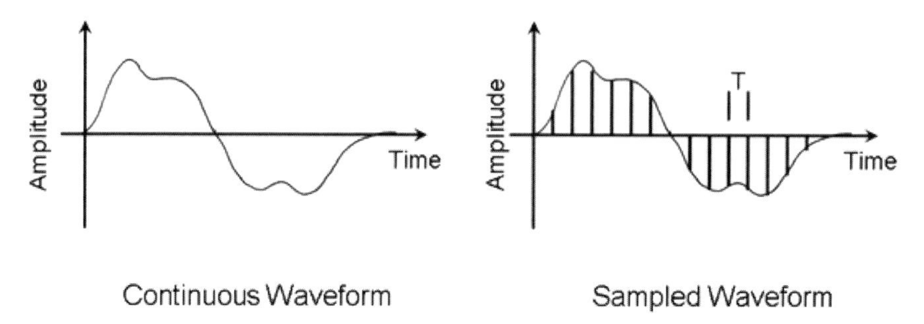

Fig. 2.10 An example of a continuous waveform is illustrated on the left, and its equivalent sampled data waveform on the right.

The problem is that digital computers can only work with finite amounts of data, i.e. sampled data (see Fig. 2.10, right side). This problem was addressed by Claude Shannon. Shannon established the basis of the information and communication revolution which has taken place over the last 60 years. One of Shannon's main contributions was his 'sampling theory', which allows data to be converted from its continuous form to its sampled form without loss of information and vice versa.

Referring to Fig. 2.10 (right) it can be seen that the sampled version of the waveform comprises vertical lines, which have discrete values and are separated by an interval T (the sampling interval). Shannon determined that all the information in a waveform can be retained in its sampled form if the original waveform is sampled with an interval T where the sampling frequency $F_s = 1/T$ and F_s is greater than or equal to twice the maximum frequency contained in the continuous waveform. This is called the Nyquist–Shannon sampling criterion. To understand what is meant by 'the maximum frequency contained in the continuous waveform' we need to introduce how a signal can be decomposed into its fundamental components. This is covered in the next section.

2.8.1 *The analysis of periodic signals*

In studying biological waveforms for the purposes of synthetic biology design it is often the case that the waveform comprises a number of components that arise from different aspects of the biological function. These fundamental components are superimposed (added) to produce the complete waveform being studied. Such an example is that shown in Fig. 2.11.

Inspection of the waveform reveals that there appear to be patterns in it; for example, similar successive peaks at approximately 0.2, 2.2 and 4.1 mins. Further inspection reveals even more patterns. However, it is difficult to proceed further just by inspection of the waveform. To precisely reveal these patterns a more detailed analysis based on Fourier series decomposition of the signal (the basis of which is covered in the next section) is very useful.

Let us now consider a second example, which is somewhat artificial but helps us to understand the decomposition of a signal into several fundamental components. Referring

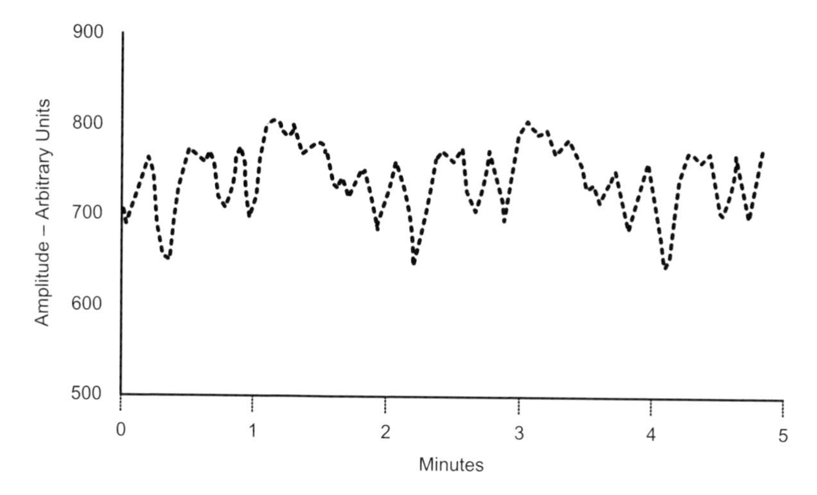

Fig. 2.11 An example of a 'continuous' biological waveform.

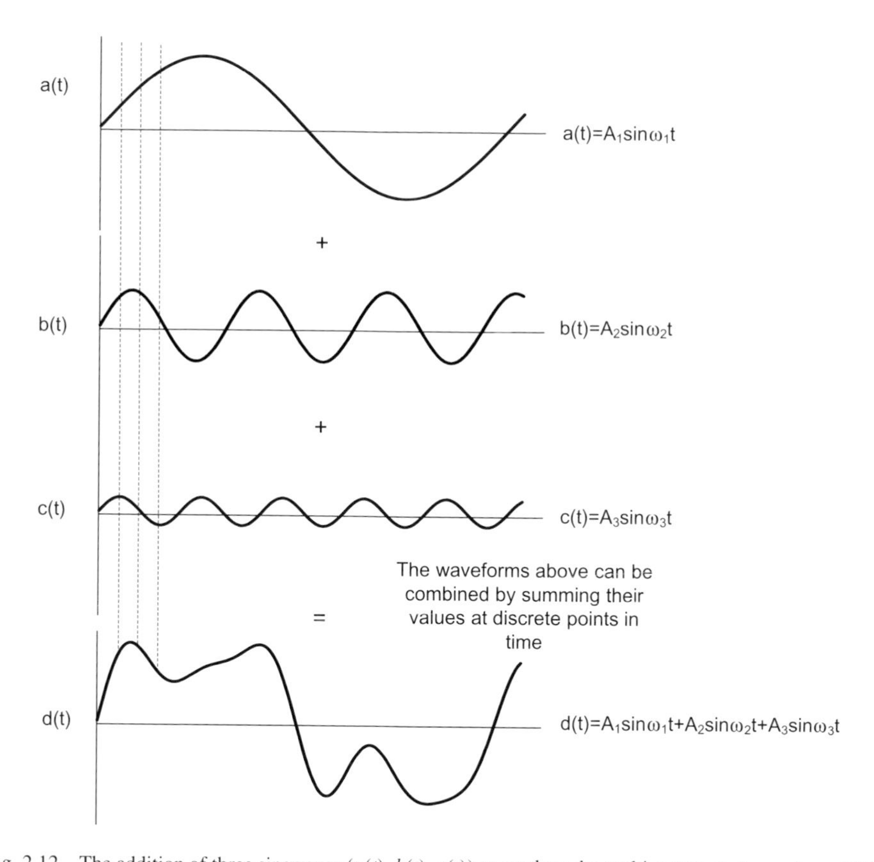

Fig. 2.12 The addition of three sinewaves ($a(t)$, $b(t)$, $c(t)$) to produce the multi-component waveform ($d(t)$).

to Fig. 2.12, consider the case of three sinewaves $a(t)$, $b(t)$ and $c(t)$, where $a(t) = A_1 \sin \omega_1 t$, $b(t) = A_2 \sin \omega_2 t$ and $c(t) = A_3 \sin \omega_3 t$. When the waveforms are summed at discrete points in time, they generate the waveform $d(t)$. This waveform can therefore be said to comprise the sum of $a(t)$, $b(t)$ and $c(t)$ — or to put it another way, the waveform $d(t)$ comprises three components $a(t)$, $b(t)$ and $c(t)$. The waveform $d(t)$ is similar to the type of waveforms obtained in experimental data (e.g. Fig. 2.11). Normally the problem in biological waveform analysis is the inverse problem, i.e. to break a waveform down into its components (in the example of Fig. 2.12, to break $d(t)$ into $a(t)$, $b(t)$ and $c(t)$).

Fortunately this is a problem that was solved by the French mathematician Joseph Fourier 200 years ago. Fourier developed a method called the Fourier transform (more on this in the next section) for breaking waveforms down into their constituent parts and to represent the components, for example, as plots of amplitude against frequency. To understand how this works, let us return to the example of the three sinewaves. Referring again to Fig. 2.12, we can see by inspection that:

$$d(t) = A_1 \sin \omega_1 t + A_2 \sin \omega_2 t + A_3 \sin \omega_3 t \quad (4)$$

where the amplitude and frequency of each sinewave is: (A_1, ω_1), (A_2, ω_2) and (A_3, ω_3). Hence, the information in $d(t)$ can be represented on a plot of amplitude against frequency, as shown below in Fig. 2.13. This plot is called the Amplitude Spectrum of $d(t)$.

2.8.2 *The time and frequency domains*

Referring to the plot of waveform $d(t)$ (Fig. 2.12), this is a plot which comprises amplitude on the vertical axis (the ordinate) and time on the horizontal axis (the abscissa). The waveform is therefore said to be in the Time Domain and the plot is called a Time Series. The amplitude spectrum of $d(t)$ comprises a plot of amplitude on the ordinate and frequency on the abscissa (Fig. 2.13). In this case, the waveform is said to be expressed in the Frequency Domain and the corresponding plot is called the amplitude spectrum. The conversion of a time series to its spectrum is a linear process and is said to be a 'conversion from the Time to the Frequency Domain'. In practice, today, the calculation of the amplitude spectrum for a particular time series is achieved via a computer algorithm called a Fast Fourier Transform or FFT (see Fig. 2.14). The FFT is a fast version of the original method developed by Joseph Fourier.

Frequently, the constituent parts of a complicated waveform can be more readily determined from its frequency spectrum. To illustrate this point, Fig. 2.15 shows the time series of a biological waveform (left), together with its frequency spectrum (right). Inspection of the time series shows that it is very difficult to determine the underlying components which comprise the waveform. However, inspection of its frequency spectrum shows that the waveform actually comprises one major periodic component at 0.05 Hz, or 20-second period, together with a range of low level noise — which impairs identification of the periodic component in the time series. This type of signal analysis by Frequency Domain inspection is a very powerful tool in the synthetic biology design toolbox.

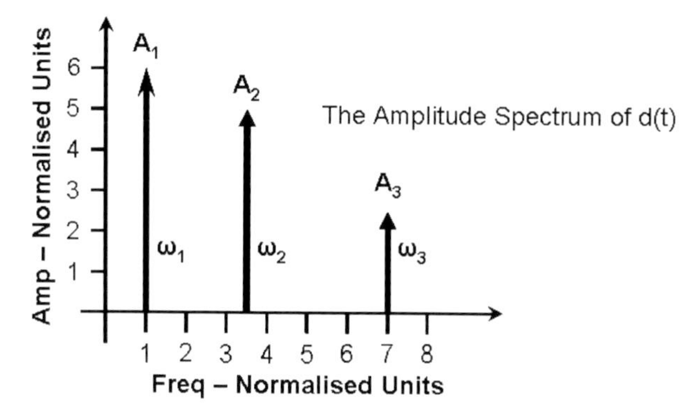

Fig. 2.13 The Amplitude Spectrum of $d(t)$.

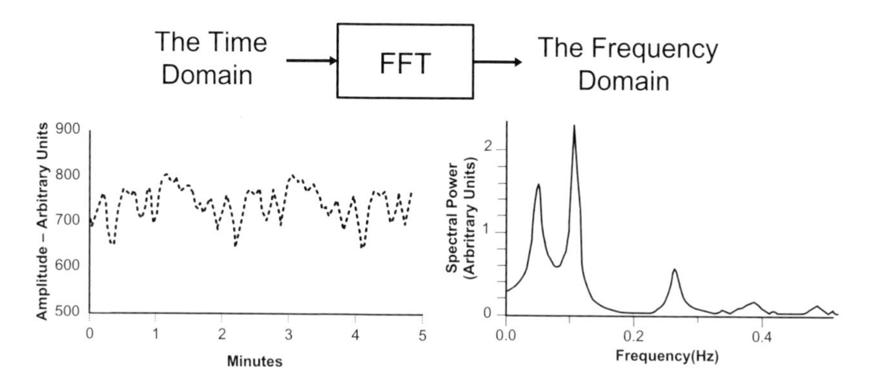

Fig. 2.14 Conversion from the Time Domain to Frequency Domain via the FFT.

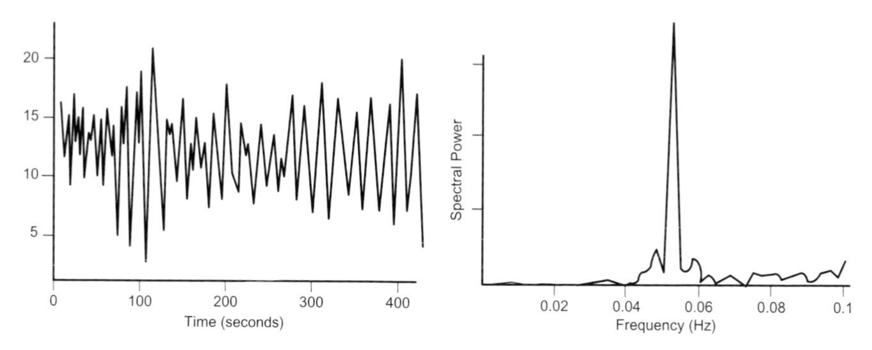

Fig. 2.15 A biological waveform and its spectrum. The waveform essentially comprises a single 20-second period component (obscured in the time domain by noise).

2.9 Systems and Control Theory

The device shown in Fig. 2.8 is a form of a dynamical system. In engineering, the analysis of such systems is achieved by Systems and Control Theory. This provides a framework to analyse properties of interconnections of input/output systems. Control Theory has been extensively used in the design of control engineering systems and in the study of physiological systems. Control Theory could also provide a systematic framework to design and build biological systems.

The challenges in modelling systems and devices are to identify and derive a suitable mathematical model, to describe the system and to determine its parameters. Modelling can be performed, for example, using a black box approach. Here the details of the system are generally ignored. The focus of the black box approach is to use the input and output of the system (or device) to define a transfer function that characterises the system behaviour. Hence, the relationship between the output variable and the input variable of the system is described by the transfer function. The transfer function is a property of a system itself, a linear system independent of the magnitude and nature of the input or driving function. This method of modelling has been widely used in the study of physiological systems. For example, for heart rate variability (HRV) the system can be described by a series of components (described in the Laplace Domain), connected in a manner that describes the overall system.

2.9.1 *Block diagrams*

The basis of the Systems and Control methodology is the representation of devices and systems in terms of block diagrams, derived from descriptions of their ODEs. As will be shown, these techniques also allow the incorporation of nonlinearities into the model.

For example, the Laplace Domain equation

$$Y(s) = G(s)X(s) \tag{5}$$

can be expressed in a block diagram form as shown below in Fig. 2.16.

For more complex systems, it is possible to break the dynamic down still further by using multiple blocks in series. So, here, $G(s)$ becomes $G_1(s)$ and $G_2(s)$ as shown in Fig. 2.17.

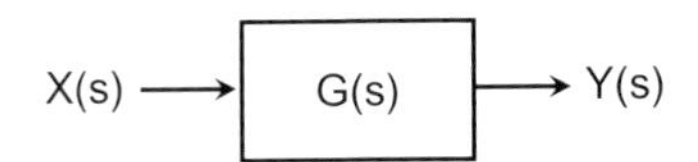

Fig. 2.16 The block diagram of version of Eq. (4).

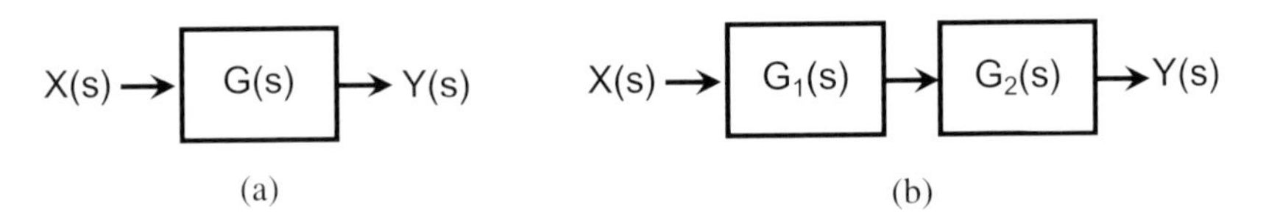

Fig. 2.17 The breaking down of G(s) into multiple blocks in series.

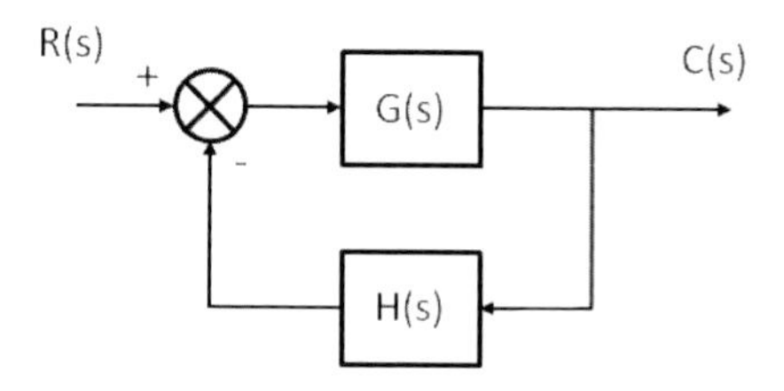

Fig. 2.18 A block diagram of a closed-loop system.

Now referring to the block diagram, the arrows only represent information flow, so that referring to the right-hand part of the diagram in Fig. 2.17, if we were to introduce an output for the block containing $G_1(s)$, say $Z(s)$, then $Z(s)$ could be defined as:

$$Z(s) = G_1(s)X(s) \tag{6}$$

The block diagram method can be extended to include feedback, as shown in Fig. 2.18. This configuration is often referred to as a closed-loop system (where the symbol with the X inside the circle is a comparator).

The method is also capable of representing blocks in parallel (e.g. if $G_1(s)$ and $G_2(s)$ were in parallel, rather than series). A final point, blocks in the Laplace Domain are said to be described by 'Transfer Functions', so G(s) actually represents a Transfer Function.

2.9.2 *Laplace transform method*

By way of example, the regulatory device shown in Fig. 2.19 can be represented as a block diagram as shown in Fig. 2.14. Block 1 represents the transcription process described by Eq. (1) and Block 2 represents the translation process described by Eq. (2).

Referring to the equations of the example device shown in Fig. 2.8, Eq. (2) is a linear equation. Hence, Laplace transformation can be applied directly. The transformation of Eq. (2) is as follows:

The Laplace transform of Eq. (2)

$$\frac{d[P]}{dt} = k_{tl}[mRNA] - d_p \cdot [P] \tag{7}$$

with zero initial conditions is given by:

$$s[P] = k_{tl}[mRNA] - d_p \cdot [P] \tag{8}$$

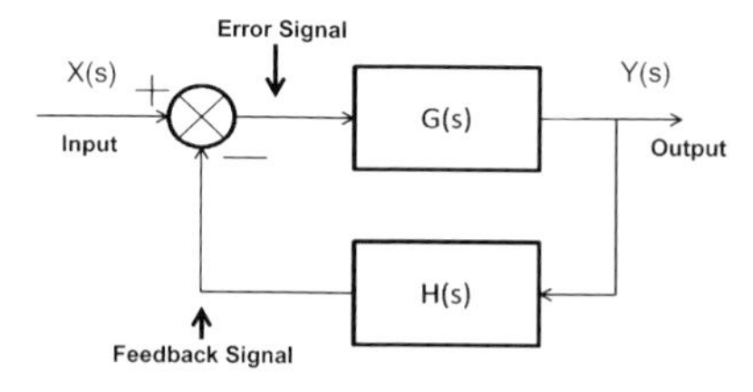

Fig. 2.19 Representation of the regulatory device using block diagrams.

By arranging Eq. (8), we obtain:

$$G_2(s) = \frac{[P]}{[mRNA]} = \frac{k_{tl}}{s + d_p} \tag{9}$$

$$G(s) = \frac{K}{s + a} \tag{10}$$

$G_2(s)$ is now in a standard form, which is called a simple pole. Simple poles have the general form where a and K are constants.

In the Laplace Domain the system solution can be represented by the transient response plus the steady state response. As stated previously, in Eq. (3), the Laplace transform operator is defined as 's'. This definition is extended to incorporate the concept of the steady state and transient responses, so s becomes:

$$s = \sigma + j\omega \tag{11}$$

where σ is the transient response and $j\omega$ the steady state response. Hence, s is replaced by $j\omega$ to obtain the steady state response of the system (usually called the frequency response).

In relation to systematic design in synthetic biology, the frequency response is often particularly useful when determining the mathematical description of parts of the device or system, in order to develop a computer model. In practice, there are a number of ways of obtaining such a description from experimental data and/or the literature. An important method of doing this is Bode Analysis.

2.10 Models and Languages

2.10.1 *Applications*

This section discusses some applications where biological analogues of electrical devices could be constructed using a synthetic biology approach. For more information on these and other devices as well as their construction from biological parts, please see Chapter 5.

2.10.1.1 *Electronics and computing*

The ability to manipulate signals and functions is a key requirement of many engineering devices and will be required of synthetic biology constructs. For example, many sensors need to combine, control or switch signals to produce a reliable output signal. With this approach the conceptual methodologies of electronics and computing can be applied to applications in synthetic biology. The operating speeds, time constants and power consumption of biologically synthesised parts and devices are likely to be very different from silicon-based electronic devices and computers. For example, biologically synthesised devices may be operationally much slower than their electronic equivalents. This may not be a disadvantage if such devices are to be used to monitor biological processes, since the time constants of the devices would match those of the environment in which they would operate. In addition, they may be driven by power supplies that derive their energy from the surrounding environment.

Biologically synthesised devices may also be capable of operating in environments that would be inhospitable to their electronic counterparts.

2.10.1.1.1 Orthogonality

One important problem which needs to be addressed is that of crosstalk in biological systems. Although biological devices are expected, conceptually, to work in a similar manner to their electronic equivalents, a key difference is that unlike electronic digital circuits the individual gates are not connected by wires. In electronic digital circuits there is no significant crosstalk, provided the circuit is properly laid out. This is not the case in synthetic biology where designed constructs sit in a melange of different substances. Since biological systems lack the same physical isolation as occurs in electronic and mechanical engineering, the interactions of biological components have to depend on the chemical specificity between them. However, the toolkit of synthetic biology now contains only a small repertoire of orthogonal regulatory elements. Here orthogonality means that the modules should not interfere with existing parts and modules in the designed biological systems as well as the genetic background circuit of the host. This constrains the development of more complicated systems that might comprise many components because the use of non-orthogonal components in one system likely leads to unintended interactions. There is a pressing need to expand the synthetic biology toolkit of available parts and modules that are truly orthogonal. This will provide the flexibility to construct larger-scale customisable systems with more complex circuitry, like the combinatorial logic circuits found in electronic engineering.

2.10.1.1.2 Logical devices and gates

The Texas Instruments data book in the 1970s transformed the application of electronic integrated circuits by allowing designers to combine logical functions using standard parts. Using a base set of Boolean functions represented by the components AND, NOT, OR, etc. any logical function can be produced, including addition and multiplication. A key aspect of this was the TTL standard permitting reliable interface between components. The development of a similar biologically engineered armoury of components performing logical functions will expand the range of biologically engineered devices that can be constructed. Logic gates are the basis of all digital devices from calculators through microprocessors to computers. The ability to design and build robust biological logic gates (BioLogic) is an important goal in synthetic biology. This is the basis upon which biological systems potentially can be controlled.

2.10.1.1.3 Timing functions and clocks

The ability to generate time delays or time intervals is another tool used in the design of electronic circuits, and will be required to provide a flexible range of functions required in synthetic biology-engineered devices. A time delay coupled with a series of inverters (NAND gates) constitute the basis of a simple oscillator which can then produce a regular clock signal.

The operation of many logical devices requires a clock to synchronise switching. Clocks can also be used with a counter to time multiple events. An example is work carried out on oscillators which are based on predator–prey dynamics using the Lotka–Volterra equations. The intention is to extend this work and look at other dynamics which could form the basis of clocks for logical devices — e.g. oscillators can now be realised in terms of protein concentrations. Combined with counters this can be used to generate a time delay, or other operations.

2.10.1.1.4 Race hazards

A race condition or race hazard is a flaw in an electronic system or process whereby the output and/or result of the process is unexpectedly and critically dependent on the sequence or timing of other events. The term originates with the idea of two signals racing each other to influence the output first, making the outcome much less predictable. In digital electronic circuits significant amounts of time and effort have been devoted to the elimination of such problems. With proper circuit layouts, digital electronic circuits are no longer subject to this problem. However, it may well not be the case in relation to biologically-based devices where the connections between gates are biochemical.

2.10.1.1.5 Counters

Once simple multi-stage logical devices have been successfully constructed, following the conclusion of the work on orthogonality and hazards, the next stage will be to construct devices which implement human-defined functions. Counters are a good example of such devices that can be used in association with clocks to define time intervals, or to trigger secondary events after a defined number of primary events. In electronic terms a counter requires an adder using the logic gates discussed above, together with a comparator that compares the count with a target number. The development of counters is a precursor to developing a range of BioLogical devices which will address issues relating to control and signalling in the context of synthetic biology.

Reading

Dorf, RC and Bishop, RH. (2008). *Modern Control Systems*, 11th Ed. Prentice Hall, Upper Saddle River, NJ.

Endy, D. (2005). Foundations for Engineering Biology. *Nature* 438: 449–453.

Franklin, GF, Powell, DJ and Workman, ML. (1997). *Digital Control of Dynamic Systems*, 3rd Ed. Prentice Hall, Upper Saddle River, NJ.

Ogata, K. (2001). *Modern Control Engineering*, 4th Ed. Prentice Hall, Upper Saddle River, NJ.

CHAPTER 3

Foundational Technologies

3.1 Introduction

In this chapter, we will examine the enabling technologies that precipitated the rise of synthetic biology and describe the foundational technologies that modern synthetic biology is built upon. Synthetic biology brings together the principles common to all branches of engineering with the sciences of biology and chemistry. We will look at how these engineering principles define synthetic biology and how advances in biology and chemistry in the last decades of the 20th century set the scene for the subject to emerge. Finally we will touch on upcoming technologies that may define the next generation of synthetic biology research.

3.2 Enabling Technologies

Modern research in biology and chemistry often overlaps and is also split into many specialist sub-divisions. To place synthetic biology in context, it is useful to start by defining some of the branches of the biosciences that provide the scientific backbone for synthetic biology and led to the enabling technologies discussed here. One of the oldest subjects in bioscience, microbiology, is the study of microbes — the single-cell organisms that include the bacteria and yeasts so often used in synthetic biology. Genetics looks at cellular function and phenotypic differences between organisms, understanding how they link to genes and genetic differences such as mutations. Biochemistry bridges biology and chemistry to give a link between biological molecules and cellular function by revealing the chemistry and structures of the cellular molecules, particularly proteins like enzymes. Molecular biology completes the circle between genetics and biochemistry, uncovering the molecules and processes in cells that take genetic information from genes and converts it into the molecules of biochemistry. Molecular biology is particularly important for synthetic biology as this subject provides the tools for engineering biology. It gives us the proteins and molecules that convert new information encoded in the DNA of genes to make proteins that change cellular function. It gives us enzymes that cut and rearrange DNA to allow new information to be put into cells and it has also given rise to the tools needed to analyse the sequences of DNA and RNA that encode all life.

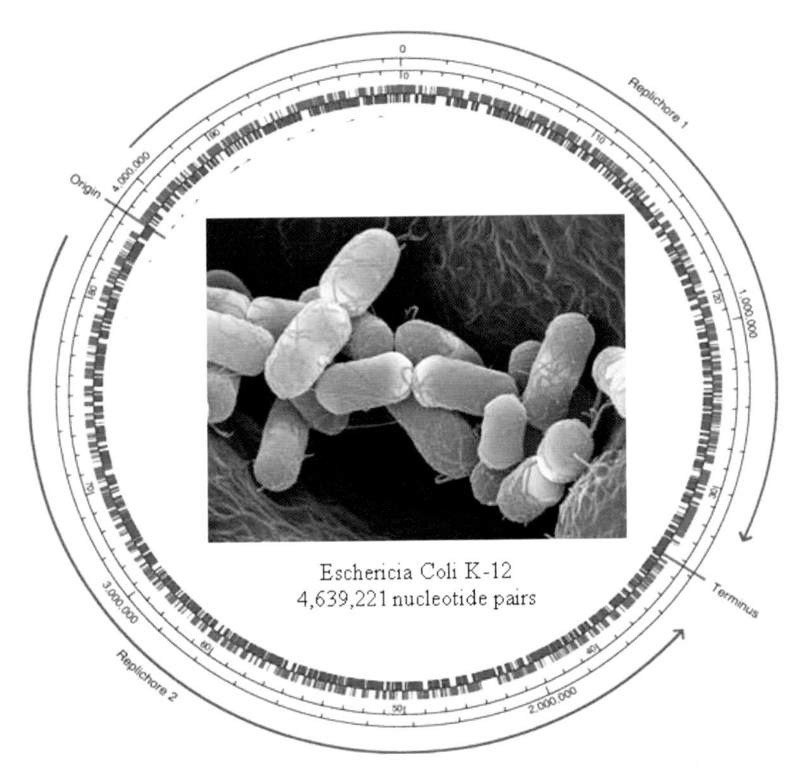

Fig. 3.1 A genetic map of the *E. coli* K-12 genome with an electron microscope image of several *E. coli* cells. (3.1(A) © Science Photo Lab. Reproduced with permission; 3.1(B) © AAAS. *Science* 277: 1453–1462, 1997. Reproduced with permission).

3.2.1 *Genome sequences*

Concurrent with the first synthetic biology research projects was the sequencing of the DNA that constitutes the human genome, a massive international project announced as essentially complete in 2003. However, it was in 1996 and 1997 that genome research made its biggest impact for synthetic biology with the sequencing of the genomes of two model organisms both used extensively in biotechnology — the bacterium *Escherichia coli* (Fig. 3.1) and the yeast *Saccharomyces cerevisiae*.

Having a complete sequence of an organism's genome is essential for understanding how a cell works and how it can be re-engineered to be a chassis system for synthetic biology. A genome sequence provides a massive source of digital information that can be used to bring together detailed descriptions of molecular biology, genetics and biochemistry — effectively being a blueprint for the function of the cell. The sudden availability of completed DNA sequences of two of the most commonly used laboratory organisms precipitated not only synthetic biology but also the tandem subject of systems biology discussed below. Parallel genome projects gave us further microbial genomes and then later the genomes of plants like *A. thaliaria* and metazoan animals like *D. melanogaster* and *C. elegans*. These genomes give molecular biologists a 'toolbox' of gene sequences to research with and it is these that provide the basis for 'parts' in synthetic biology.

3.2.2 *Open online databases*

The World Wide Web has had a big impact on research science, allowing scientists to share discoveries in faster and more information-rich ways than the printed publication methods, which dominated until the 1990s. Online databases have been particularly important in enabling technology development for synthetic biology. The ability to share graphical representations of large data sets is crucial to modern science. An example of this is the genome projects discussed above. These projects output millions of bases of DNA sequence that are too large to be disseminated on paper, but are perfectly suited to online databases. The Saccharomyces Genome Database (SGD) is an excellent example of this, providing the home to 12 million base pairs of yeast DNA sequence and the annotation of over 6,000 genes and thousands of experiments. As with many other online databases, such as FlyBase for Drosophila fruit fly researchers, the presence of such an essential website for research in this area also provides a focal point for research communities, and with freely available online data comes open-source sharing. A common model for an open online database is as a community wiki, where researchers can input and edit information relevant to their field of research and freely see that of others. There are now hundreds of open online databases containing biochemical, genetic and molecular biology information for a variety of organisms. Synthetic biology is able to make use of many of these databases. Databases that describe chassis cells, typically *E. coli* or *S. cerevisiae*, provide information on how to re-engineer the existing cellular systems. Databases that list enzymes and metabolic pathways from across nature are used to design systems that add 'foreign' genes to give new biological functions. Finally, synthetic biology has established its own online databases to help share information on experiments and designs. Two examples, both of which are wikis, are OpenWetWare.org, which is used to share experimental protocol information, and partsregistry.org, which holds vital information on standard biological parts discussed below and in other chapters.

3.2.3 *DNA sequencing*

DNA sequencing is one of the main tools of modern biosciences. Early molecular biology research and the first genome projects used traditional Sanger sequencing to sequence individual DNA pieces up to 1,000 base pairs at a time, and relied on gel electrophoresis and fluorescent dye chemistry to get the base pair sequence of DNA. For much longer sequences, such as whole chromosomes, a strategy called shotgun sequencing was employed in the 1990s by J. Craig Venter and others. In this technique a large piece of DNA is cut into many smaller fragments and sequenced by Sanger sequencing in parallel several times. Overlapping DNA data from each sequencing read is fed to an algorithm that re-assembles the sequence of the large DNA piece *in silico*. Parallel sequencing played a particularly large role in enabling rapid and affordable sequencing of bacterial genomes and is still the primary strategy to sequence chromosome-sized DNA. The strategy suits DNA sequencing on an industrial scale and has led to the creation of specialised 24-hour sequencing centres with automation, and a continual, exponential drop in the price and time it takes for researchers to have cloned DNA sequenced. For synthetic biology, verification of the DNA sequence of novel constructs is crucial and so having fast, cheap DNA sequencing is a prerequisite for projects of every size. In particular, DNA sequencing is essential for projects where devices are hosted in bacteria as the fast evolution within these organisms can quickly add mutations to sequences. The technology behind the physical sequencing of DNA has moved from gel electrophoresis to capillary electrophoresis for individual cloned DNA sequences common in synthetic biology. For genome-scale projects the sequencing chemistry has completely moved on to what are referred to as second- and third-generation sequencing technologies that sequence short DNA fragments in a massively parallel manner. These technologies create huge datasets and rely heavily on computation and algorithms to determine the genome sequences.

3.2.4 *DNA synthesis*

The technology directly related to DNA sequencing is DNA synthesis. DNA synthesis broadly falls into two categories: (i) oligonucleotide synthesis and (ii) gene synthesis — the latter of which is discussed below. Oligonucleotide synthesis simply means the custom synthesis of short, single-stranded DNA oligonucleotide sequences, usually to be used as primers for PCR amplification of gene sequences. A customer defines the DNA sequence that they would like and simply sends the base sequence of their order to a synthesis facility, usually a company. Oligonucleotide synthesis is a chemical technology — a solid-phase synthesis performed using phosphoramidite chemistry to link A, C, G and T building blocks into chains as long as 150 nucleotides. As with DNA sequencing, oligonucleotide synthesis is typically done on an industrial scale at automated 24-hour centres. Process advances and commercialisation have led to a continual decrease in cost and turnover times for synthesised DNA oligonucleotides that is beginning to match those for DNA sequencing.

The decrease in costs of oligonucleotide synthesis and DNA sequencing has been overtaken by an even more rapid and exponential decrease in the cost of gene synthesis (Fig. 3.2). Gene synthesis is the custom ordering of commercially produced double-stranded sections of DNA, usually from 200 to over 2,000 base pairs in length. Gene synthesis costs are dependent on oligonucleotide and sequencing costs as the custom ordered genes are assembled from pools of synthesised oligonucleotides and then sequence-verified before delivery to the customer. Gene synthesis first requires optimised design and then synthesis of the required oligonucleotides, which are typically 40 to 60 bases in length. These oligos make up both the sense and antisense strands of the duplex DNA product and contain substantial overlap (Fig. 3.3). The gene is assembled from the oligonucleotides by allowing overlapping sequences to anneal and then using purified ligase enzymes and/or DNA repair enzymes to seal gaps. Alternatively, PCR amplification methods can be used; two of the most common are called progressive overlap extension PCR and thermodynamically balanced inside-out PCR. Assembled synthetic genes are purified, sequence-verified and regularly

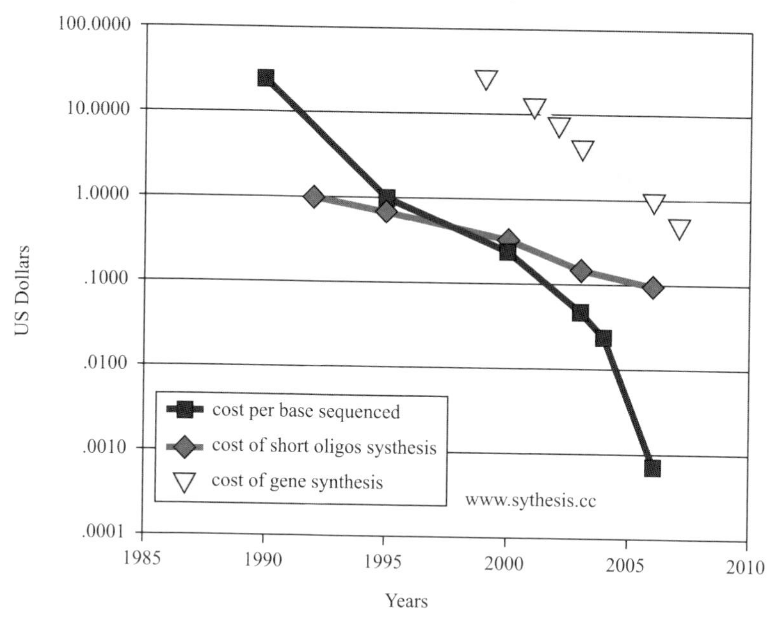

Fig. 3.2 The 'Carlson Curve' — the fall in DNA sequencing and DNA synthesis costs plotted over the last two decades. Note that cost is on a log-scale, so a straight line decrease is actually an exponential decrease. (Courtesy: Rob Carlson — www.synthesis.cc).

require error-correction using DNA repair enzymes, usually due to problems in the quality of the oligonucleotides used for assembly.

A strong link between synthetic biology and gene synthesis has always existed. Gene synthesis is a short cut to routine laboratory work required for molecular biology such as repetitive cloning and PCR. Gene synthesis allows DNA sequences placed online by researchers to be converted back into DNA constructs in other labs around the world, accelerating the sharing of resources. Importantly, gene synthesis also allows the production of entirely new synthetic gene sequences that have never existed before in nature. For synthetic biology, not only does this offer the freedom to produce novel designs for biomolecules like proteins or regulatory DNA sequences, but crucially allows protein-coding sequences to be codon optimised. Codon optimisation is desirable (and sometimes a necessity) when DNA sequences encoding proteins are transferred from diverse organisms into microbial cells. Throughout nature there is large variation in which codons are preferred by each organism and so the native DNA sequence of, say, a tobacco plant enzyme may be very poorly converted into protein when placed in *E. coli*. This is because codons that are highly efficient in plants may be rarely used in the bacteria (see Chapter 1). Rare codons in a gene cause the host cell machinery to stall when making the protein as the tRNAs required to match these codons are scarce. In order to allow synthetic biology to engineer complex devices and pathways using the biochemical diversity found across

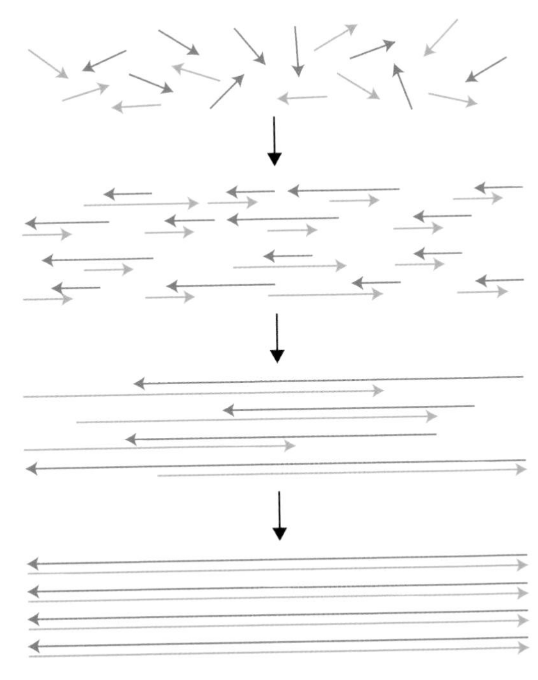

Fig. 3.3 Diagram of gene synthesis from overlapping oligos by assembly PCR.

nature, codon optimisation through gene synthesis is invaluable. Gene synthesis with codon optimisation has the further advantage of allowing genes to be re-coded at the DNA sequence to specifically contain or be absent of defined sequences, such as those sites recognised by DNA-editing enzymes like the restriction enzymes that play a key role in synthetic biology DNA assembly.

3.2.5 *Systems biology*

A genome sequence is an example of a data-rich experimental project that requires online databases to disseminate results. While genomics was one of the first areas in bioscience where this occurred, many modern experimental projects now collect millions of data points through the use of what is called 'high-throughput techniques'. In particular, experiments that take parallel measurements of molecules in cells produce very large datasets. An established example is DNA microarrays, which are used to quantify the level of all mRNAs in a cell at a given time, and so give a picture of what genes are being expressed to make protein and by how much. Before genomics and DNA microarrays, experimental biological data had been a precious resource established by individual researchers. However, now the amount of experimental data produced by some techniques has surpassed what classical research methods can adequately describe. To address the gap between data and interpretation, researchers have taken advantage of the increasing availability of computer power to use mathematical-based analysis of data to infer cell function. By writing *in silico* mechanistic models of how cells may operate using mathematical software it has become possible to systematically analyse and integrate large complex datasets to derive how cells respond to stimuli and the nature of the processes that control their life cycle. This research approach, which began after the first genome sequences were completed, is known as systems biology. It is closely associated with synthetic biology because both disciplines use mathematical models to describe and predict function. For systems biology, computation is used to dissect a cell's natural behaviour from wide-ranging experimental datasets. In contrast, modelling is used for synthetic biology, as it is in engineering, to aid in the design and description of newly constructed functions (see Chapter 6). A useful way to compare systems and synthetic biology is to think of how they approach working with cells and data. Systems biology is considered 'top-down', as measurements of outcomes are used to build a model of the inner workings of the cell as a system; synthetic biology is 'bottom-up', as the inner workings are measured and then put together and built to create outcomes. The two research fields complement one another in methodology, as modelling strategies such as ordinary differential equations and stochastic simulations are common to both fields, and the initial use of mathematics to describe cellular processes by systems biology has paved the way for its evident application to synthetic biology.

3.3 Foundations

3.3.1 *Standard DNA assembly*

The foundational technologies that underpin synthetic biology and differentiate it from traditional DNA manipulation by molecular biology are based on engineering. In other engineering disciplines like electronic and mechanical engineering, standards are an essential foundation, allowing parts and data to be shared worldwide. In particular, the standards that define how parts are linked together are important for giving modular composability, and examples of this can be seen all around us, like the USB standard for linking electronic devices and standardised threading on screws. To facilitate modular composability in engineering biology and ease the re-use of defined DNA parts, synthetic biology has built a foundation of standard DNA assembly that allows parts to be sequentially combined into working devices, pathways and systems.

Biological devices assembled in the initial projects of synthetic biology used unique DNA sequences recognised and cut by restriction enzymes. These sites, when placed at the boundaries of DNA parts, allowed them to be linked together and enabled parts to be removed and/or replaced as necessary using various DNA-editing enzymes to cut and paste. The modularity of this approach is evident, yet it requires a long list of available unique sites and these have to be added accordingly to the ends of parts by DNA-modification techniques like PCR. To overcome these limitations the BioBricks™ assembly standard was developed by Tom Knight at MIT in 2003. The BioBricks™ standard places the recognition sites of three specific restriction enzymes at each end of every part, and digestion and ligation at these sites allows parts to be assembled together in a standard fashion. Importantly, the use of different restriction enzymes whose digested recognition sites can still ligate together (isocaudomers) means that standard parts can be linked together in a manner that removes the digestion sites from the part boundaries in the composed product. Thus any product of standard assembly of BioBrick™ parts becomes itself a usable, larger composite BioBrick™. For a full description of BioBrick™ parts and their assembly, see Appendix 1.

BioBricks™ assembly, however, has limitations that pose a challenge in designing and engineering synthetic systems. Repetitive 'scar' sequences are left behind at part boundaries which can affect the function of the neighbouring parts, and parts cannot be used which contain within them any of the restriction enzyme recognition sequences that are used for standard assembly. These limitations along with the slow methodology of BioBricks™ assembly means that few of the major research projects in synthetic biology use this assembly method, instead relying on numerous bespoke methodologies. However, the foundation that standard assembly provides is clear as it allows any researcher with the same set of restriction enzymes to reliably assemble thousands of prototype devices from standard biological parts. It is a relatively easy-to-use method and this is evident annually at the International Genetically Engineered Machine (iGEM) competition where undergraduates describe complex devices assembled and tested in just three months (see Chapter 8). In effect, it is a useful prototyping tool analogous to the plastic models that architects produce before buildings are constructed. This reliable method of construction allows prototypes to be tested that can later be replaced by tuned, professional final products, and experience with BioBricks™ assembly paves the way for researchers to tackle complex projects in synthetic biology.

3.3.2 *Standard measurement*

Regardless of the method of assembly of biological parts, it is important to have standard measurement of their function and efficiency. For synthetic biology to be an engineering discipline, DNA parts of all types, whether in BioBricks™ form or not, need to be adequately described by data. This allows parts to be re-used by other researchers without themselves having to replicate the most basic experiments. It also enables model-based design and description of devices and systems as the characterisation data of parts helps build the mathematical model, defining the working parameters that give realistic output. In other engineering disciplines, having adequate characterisation data available for each part is essential. For integrated circuit engineers in the 1970s and 1980s, the TTL Data Book described the necessary characteristics of transistors and resistors required for hobbyists and professionals to assemble complex digital circuits. An equivalent to this data book will be required for biological parts as a foundation for synthetic biology to build complex systems.

Precise, quantitative measurement of biological function has always been a challenge. New technologies are able to increase the speed and accuracy of measurements from biological systems and give exponentially more data. For synthetic biology, the key foundational technologies used in the standard measurement of parts are microscopy, flow cytometry, colorimetry, luminometry and fluorimetry. This is because the majority of synthetic biology experiments use reporter systems to measure the functional output of a DNA part, device or system. The most accurate and easy-to-use reporter systems rely on accurate detection of cells changing colour or producing fluorescent molecules (fluorophores) or luminescent molecules (luminescence). The most commonly used reporter system for standard measurements in synthetic biology is Green Fluorescent Protein (GFP). This is a protein originally isolated from the jellyfish *Aequorea victoria* that exhibits bright green fluorescence when exposed to blue light. The DNA sequence encoding this protein can easily be placed into other organisms to make them fluoresce green and fusing the GFP-encoding DNA sequence to the coding sequence for another protein effectively tags that protein with green fluorescence. Output of biological function can be measured using GFP by fusing its encoding DNA to proteins whose expression levels or location need to be determined. Alternatively, directly placing the GFP DNA sequence alongside regulatory DNA sequences like promoters enables researchers to measure how those sequences modulate gene expression levels. GFP levels are quantified at the population level by measuring the green fluorescence output of a volume of cells using fluorimetry. Quantification can also be performed at high-speed for single cells using flow cytometry which passes individual cells through a microfluidics chamber past a blue laser and then detects green output. Finally, fluorescence microscopy can be used to track and (using image analysis) quantify GFP expression within single cells and for populations of cells.

Alternative reporter systems to GFP include natural enzymes such as β-galactosidase and catechol 2,3-dioxygenase that catalyse quantifiable colour-change reactions when specific substrate chemicals are provided. Luminescence is also used as a reporter mechanism, taking natural luciferase enzyme genes that exist in fireflies and jellyfish and expressing these in engineered cells to produce a quantifiable light output. There also now exists a whole range of different fluorescent proteins other than GFP — including blue, cyan, yellow and several shades of red fluorescent proteins along with GFP variants optimised for a wide variety of different cellular conditions (Fig. 3.4). Research and development in biotechnology, using mutation and selection methods, has enabled scientists to engineer new properties into natural proteins and discover new variants of fluorescence. The many types of available reporter systems, particularly fluorescent proteins emitting different colours, allow researchers to measure the output of several parts, devices or systems at once. For example, the gene expression output of two promoters in an engineered yeast cell can be quantified at the same time by placing the GFP DNA sequence downstream of one promoter and an RFP (red fluorescent protein) DNA sequence downstream of the second promoter. Using a fluorescence microscope or a flow cytometer with the correct detection filters and excitation lasers, the researcher can simultaneously record green and red fluorescence, and in doing so infer the output efficiency of the two promoter sequences.

Although technology is improving, the precision of measurements performed by different research groups often produces data that shows large differences. This is due to a variety of reasons. Firstly, biology is inherently 'noisy' — within an apparently identical population of a million bacteria cells some will exist that are at different stages in their life cycle, different sizes and with varying access to nutrients and oxygen. Even at the molecular level within the cell there are events like transcription that are stochastic in their

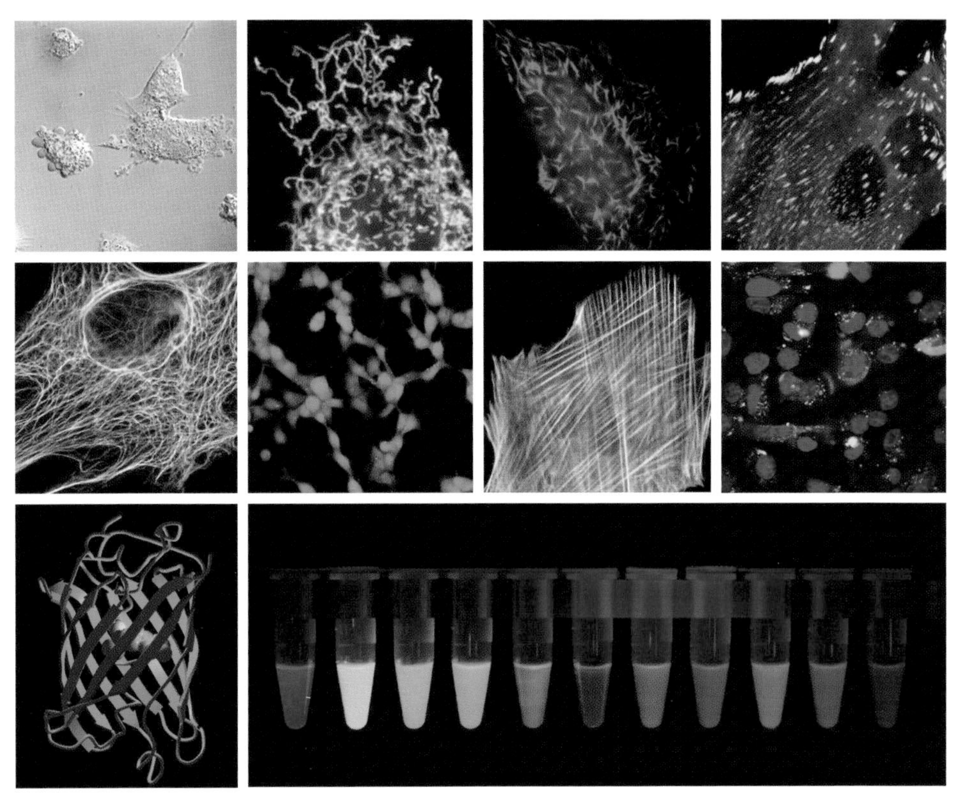

Fig. 3.4 Fluorescent proteins are available in many shades of colour, thanks to natural diversity and extensive in-lab evolution experiments. (© Paul Steinbach, Roger Tsien Laboratory. Reproduced with permission).

nature. Secondly, cells used by different researchers can be quite different — for example two synthetic biologists engineering an *E. coli* chassis could be working with different strains of the same organism, which have small genetic differences that cause significant changes in their behaviour. Even when two laboratories both use the same strain, the ability of bacteria to quickly acquire mutations means that the two labs may still not have genetically identical systems. Finally, the largest source of variation in measurement comes from the researchers themselves, who set up their experiments differently, following different protocols, have different lab conditions such as temperature and air pressure and use different equipment from a variety of manufacturers to make their measurements.

Although it may be possible to limit this variation by vigilant standardisation of cell strains, chemical reagents, equipment and experimental protocols, it is also possible to do so by using reference elements to aid in standard measurement. A good example of a reference element is the metre. The metre is the standard unit used to measure distance and is based on a length chosen by scientists and engineers. To quantify any distance, an engineer will directly compare that distance to their own measurement tool standardised to the official metre. Just as in measuring distances in engineering, in synthetic biology it has been shown that using a reference element can help standardise measurement between research groups around the world. In one example of this a promoter sequence active in growing *E. coli* was set as the benchmark reference element for promoter efficiency. Quantifying the green fluorescence output of cells engineered to express GFP under the control of this promoter gives each lab their reference output measurement. Labs can then compare the measured output of other promoter sequences to this reference measure and share their data as relative expression output compared to the standard that each lab has quantified. As a foundational technology for synthetic biology, standard measurement using reference elements enables more accurate sharing of characterisation data among the community.

3.3.3 *Abstraction*

Natural DNA sequences often encode multiple overlapping functions, providing a challenge for synthetic biology. For example, the initial bases of an mRNA sequence can affect the efficiency of translation but they can also modulate the mRNA lifetime, and are coded for by DNA bases that themselves are usually part of the promoter sequence. Overcoming this and defining each part and its function is a major task in synthetic biology, but it is essential for providing the modular composability that will allow the engineering of novel biological devices and systems.

Composability is a fundamental part of what is known as abstraction in engineering, where intimate understanding at one level (e.g. parts) is not required to work at another level (e.g. devices). Abstraction allows the construction of complex devices without the full understanding of the processes at each scale (Fig. 3.5). For example, in the computer industry,

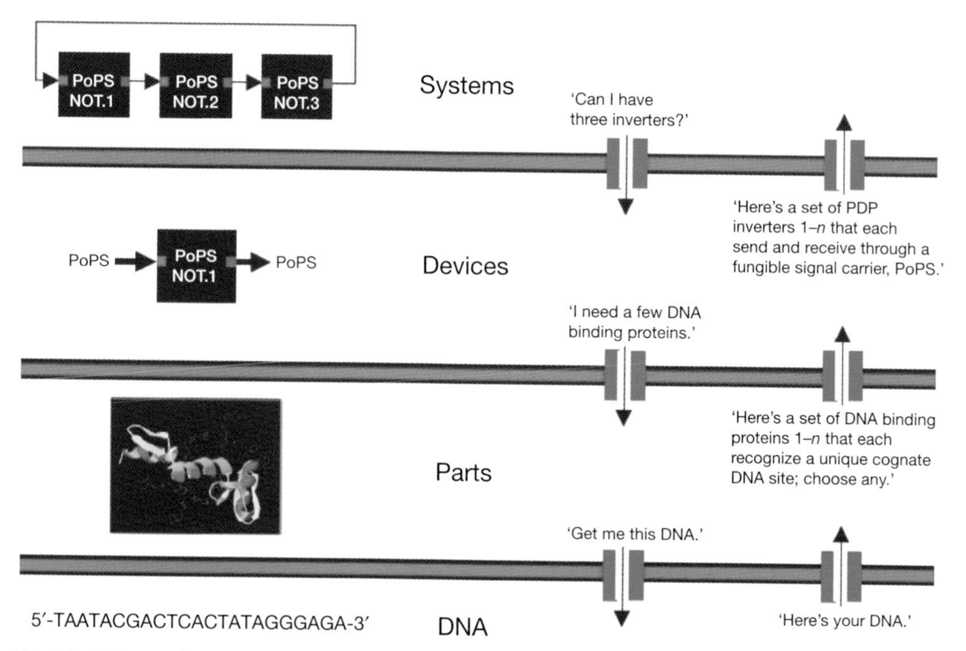

Fig. 3.5 Diagram demonstrating the different abstraction levels in synthetic biology. (Courtesy: *Nature* 438, 24 November 2005, 449–453. Reproduced with permission).

a laptop is designed and assembled by people who have no expertise in how transistors are fabricated into microchips. In synthetic biology, abstraction allows there to be experts on parts like promoters and transcription factors who can characterise these standard parts and pass them over to specialists who design and build devices or engineer pathways. These specialists can then, in turn, offer their expertise to systems-level designers who can work with little or no knowledge of how parts in their designs operate at the molecular level. By having abstraction in synthetic biology it allows complexity to be engineered without a reliance on highly trained experts who understand cellular biology at many levels of detail. In order to enable abstraction it is essential that biology is engineered in a modular fashion with defined parts. These parts are required to be measured with a suitable degree of precision so that their characterisation data can be used to accurately design and build devices and, in turn, these devices themselves need to be adequately characterised and modular for them to be interfaced into synthetic systems.

3.3.4 *Modelling*

As in all aspects of engineering, mathematical descriptions of designs and processes are an essential tool in synthetic biology and provide a foundation for robust and scalable construction of complex devices and systems. Mathematical models are applied in synthetic biology both before a device or system is assembled and characterised in cells, and also afterwards. Modelling prior to building and testing a device is typically intended to be predictive. That is to assess the likelihood that a design will work, to suggest other possible designs, to simulate the potential performance and to derive the parameters that will give the intended robust function. Deriving the optimal parameters by predictive modelling allows the researchers to select the appropriate parts that should build a successful device or system. For example, when constructing a regulatory gene network, inputs and outputs at logic gates need to be tuned appropriately; parameter estimation informs the relative expression levels required for the regulatory proteins and by selecting promoter parts of different strengths to control expression of these proteins, researchers can build balanced and predictable devices (Fig. 3.6). The use of modelling after the assembly and characterisation of a biological device or system serves several other useful roles in synthetic biology:

(i) Modelling can be used to give a fundamental description of a novel process, explaining the characterisation data and providing a mathematical foundation for the synthetic design. A model of a characterised synthetic device or system can often then be used to derive a mathematical understanding of an equivalent natural process.
(ii) A model describing the experimental data can be used typically by simulation, to quickly explore how a design performs in a wide variety of other potential experimental conditions, including ones which would be difficult to examine in a typical laboratory.
(iii) A post-testing model provides the crucial link to the second round of an engineering cycle, allowing the data derived from testing an initial prototype in a first cycle to direct the design and construction of one or more optimised second-round designs.

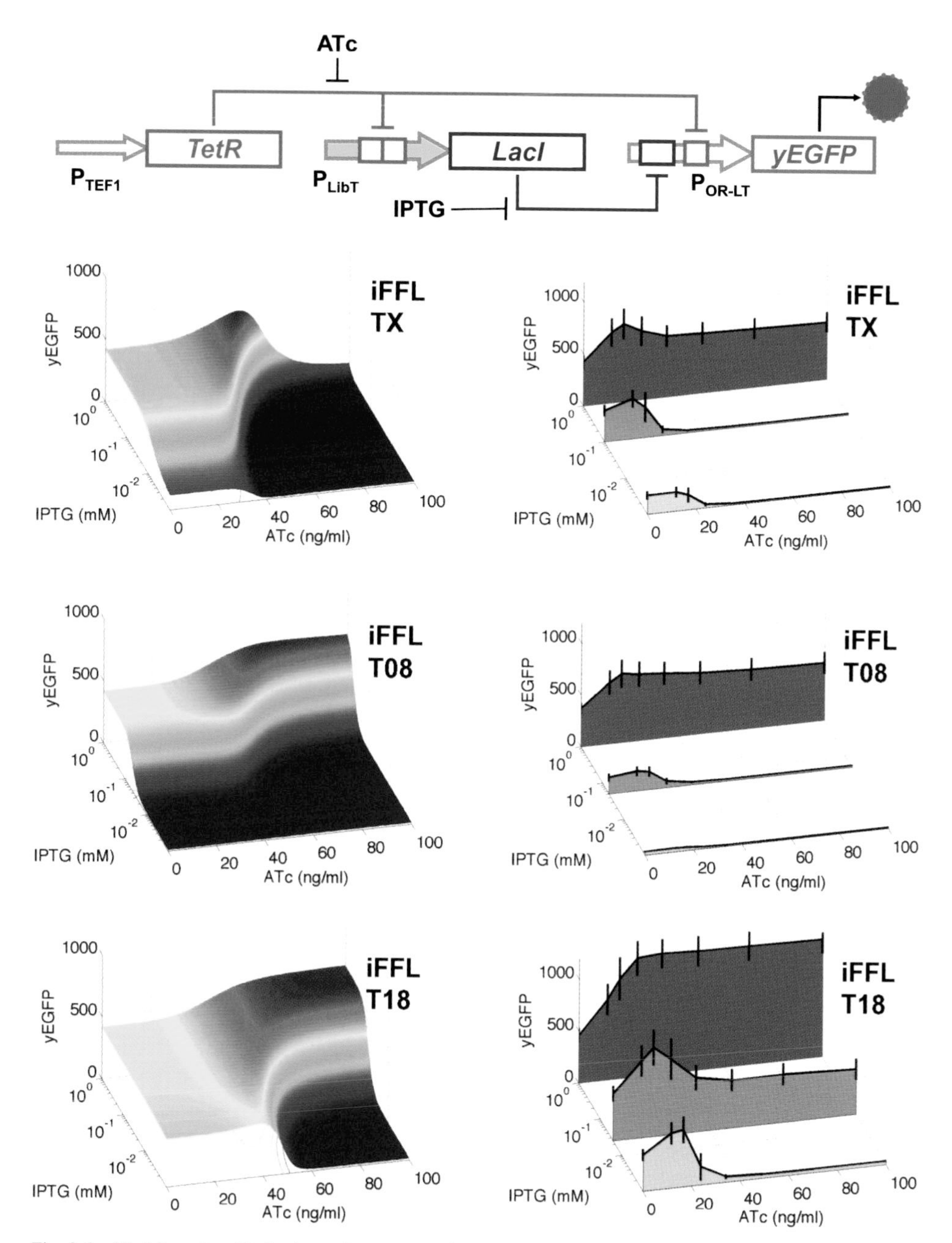

Fig. 3.6 Model-predicted behaviour of gene networks can be used to inform how a network will perform when DNA parts from part libraries are interchanged. (Courtesy: *Nature Biotechnology* 27, 2009, 465–471. Reproduced with permission).

Modelling in synthetic biology is important at all scales of work, from describing how protein parts bind their substrates or predicting the folding of RNA motifs, up to analysing how the flux of metabolites runs through enzymatic pathways and understanding how cells interact with one another in engineered consortia. Models of all scales can be done using a variety of techniques including ordinary differential equations (ODEs), stochastic differential equations (SDEs), Master equations and Boolean networks. A detailed description of modelling for synthetic biology is found in Chapter 6. A foundation for modelling in synthetic biology is intended to allow abstraction and to increase the speed and reliability of building synthetic biological devices and systems. Experimental molecular biology is, by comparison, time-consuming, expensive and can be unpredictable. Computational design and simulation is cheap and rapid. By placing more of the process of the synthetic biology engineering cycle *in silico* there is accelerated innovation and application at lower costs.

3.3.5 *Parts registries*

Science relies on the dissemination of data and the exchange of materials such as DNA and cells from one group to another. Traditionally, this was done through collaboration and via conferences but with increased volume of experimental data collected, open online databases have become more important and also provide a home for valuable basic information about experiments, such as the methods used and the raw data collected. With the decreasing cost of gene synthesis, online databases are also a way to exchange materials, as posting the digital information of a DNA sequence allows others to have this made.

For synthetic biology, sharing data and exchanging DNA sequences is particularly important as it allows engineered systems to be rapidly re-used and modified by others. This is especially true for engineering that uses standard parts to compose complex modular designs. The same community of synthetic biologists that developed the BioBrick™ and iGEM also co-developed a parts registry for standard BioBricks™ parts. The Registry of Standard Biological Parts (partsregistry.org) is maintained in the USA by a dedicated team and provides a wiki-like environment for iGEM participants and other BioBricks™ users to upload DNA sequence information and characterisation data for parts, devices and systems they have constructed or have re-used. The registry and its role with iGEM are both discussed further in Chapter 8. For BioBricks™, this registry is an important resource. It gives a full catalogue of available parts that is easily accessible. It provides some data describing their function, performance, design, use and reliability. This information, collated at a single website, enables the synthetic biologist to model, design and build without having to search through thousands of research publications and request materials from hundreds of researchers. A parts registry instead offers a real catalogue of all the usable components in one place and in doing so sets standards for how these should be designed and described. By setting these standards and allowing open access for a participating community, a parts registry offers an essential foundation for engineering biology.

The BioBricks™ parts registry has catalogued several thousand parts, mostly through the iGEM competition. It provides a major resource for synthetic biology, but is sometimes lacking in usable characterisation data for many parts due to it primarily being part of an undergraduate student competition. As companion databases, synthetic biology researchers have developed professional parts registries which are not limited to BioBricks™ parts. These are designed to contain rigorous description and characterisation data for each biological part. This makes modelling and design more predictive and enables rapid construction of working devices and systems. Closed-source parts registries of this kind exist at commercial synthetic biology companies and an open professional parts registry was founded in 2009. This open-source parts registry, The International Open Facility Advancing Biotechnology (BIOFAB), is a USA-based public-funded facility collaborating with synthetic biology research centres around the world, including Imperial College London. BIOFAB and collaborators produce and accurately characterise thousands of high quality standard parts to populate a professional registry that is open to the public (Fig. 3.7). The BIOFAB registry is intended to provide a community foundation for professional synthetic biology and industrial engineering of biology. BIOFAB projects define the standards for measurement and provide the necessary part information required for the design and modelling of new devices and systems. The commitment of both the BioBrick™ Registry and BIOFAB to be open source is also an important factor and is described further in Chapter 9.

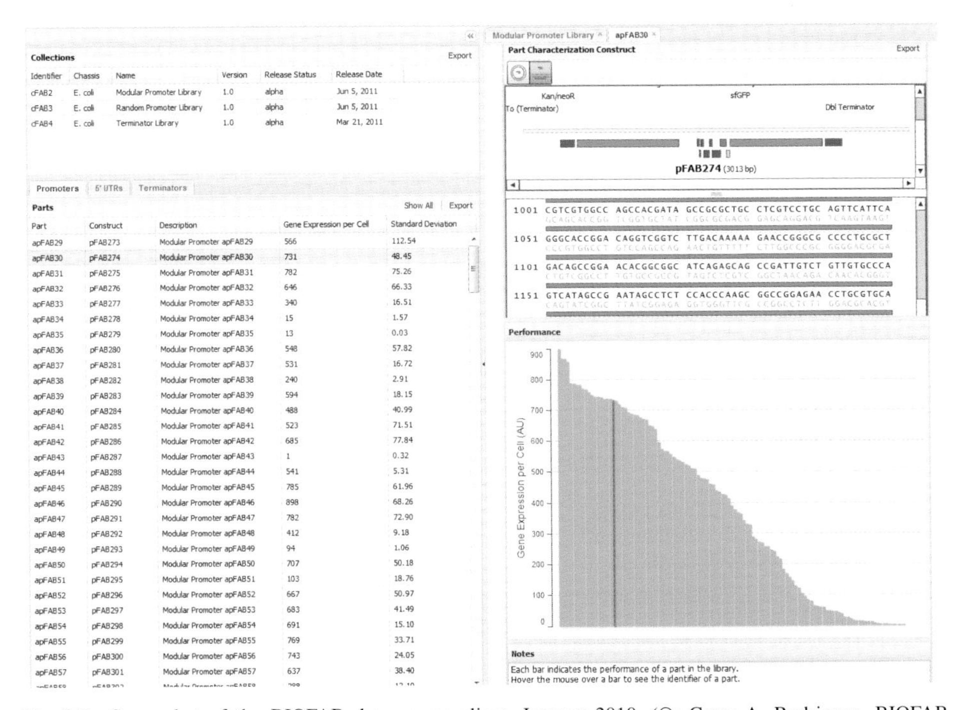

Fig. 3.7 Screenshot of the BIOFAB data access client, January 2010. (©: Cesar A. Rodriguez, BIOFAB, http://biofab.jbei.org/. Reproduced with permission).

3.4 Upcoming Technologies

Synthetic biology has put into place foundations for engineering biology based on the technological advances of genomics, bioinformatics, DNA sequencing and synthesis. The future for synthetic biology is hard to predict but is likely to be shaped by new technologies and goals that are being set by the community of synthetic biology researchers. Ultimately, the main aim of synthetic biology is to reduce to a minimum both the experimental laboratory work and the scientific enquiry of the discipline, and instead turn it into a predictable technology suitable for systematic biological design and industrialisation. This is not to say that applying synthetic biology approaches will also uncover significant new and fundamental insights into biological systems. The drive to achieve both of these aims is notable in several technologies now impacting on synthetic biology.

The rise in the availability of high-power computation will enable more complex mathematical modelling, and in particular will allow simulations of systems that are based on combinations of different models, each working on different scales (multi-scale modelling). Combining complex models and simulations with improved part characterisation data will accelerate the use of computation in predicting how novel synthetic designs will perform, and then how to optimise them further. The end goal for modelling in synthetic biology will be the development of Computer-Aided Design (CAD) software (BioCAD) environments, accessible to researchers who work at the different levels of abstraction. In such a program, DNA sequences of parts or devices are treated as modules that can be virtually assembled in a 'drag-and-drop' environment. The functioning of the virtual assemblies in a variety of conditions can be simulated until an optimal design is finalised, upon which the parts list and full DNA sequence to make the design is provided along with detailed instructions on how a laboratory researcher or robotic platform could assemble and test it. CAD software like this is common in other engineering disciplines such as microchip fabrication, and some basic CAD software for synthetic biology is already becoming available. An example of such a CAD program (ClothoCAD) is provided in Fig. 3.8.

Computation can also provide an alternative to parts registries in synthetic biology. The function and efficiency of some biological parts can be linked directly to the DNA sequence that encodes them by sequence-to-function models. In 2009, an example of this was shown when a simple RNA part — the ribosome binding site (RBS) — was modelled with enough detail to allow a predictive tool to be produced. This software tool, available online, provides the user with the DNA sequence that they can have synthesised to produce an RBS part with a desired function and efficiency. With cheap custom DNA synthesis and predictive tools for custom designing of parts, it is possible to use modelling to design some simpler parts *de novo*, instead of obtaining existing parts from registries.

Gene synthesis will continue to play a large role in synthetic biology. Advances in the understanding of codon use in various organisms are allowing genes to be designed with increasingly reliable and predictable levels of expression. The speed at which long DNA sequences can be produced by commercial companies is continually increasing and the cost continues to fall. Recently described methods for genes synthesis from microarray DNA promise to drive costs down considerably further, as the parallel production of thousands of

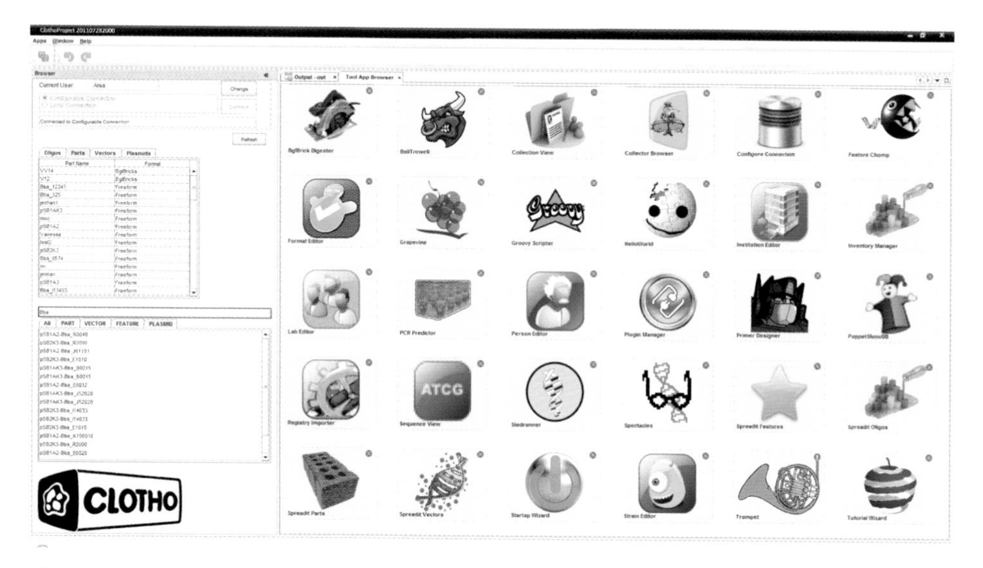

Fig. 3.8 Screenshot of the Clotho CAD tool, January 2010 ©: Doug Densmore, Clotho, http://www.clothocad.org/. Reproduced with permission).

DNA oligonucleotides on microarrays is much cheaper than traditional chemical synthesis. This is an example of how cost can be decreased while output is increased by using miniaturisation of typical laboratory processes. Miniaturisation is expected to help advance synthetic biology and is already a strategy to improve the characterisation of parts and devices. This is done using microfluidics — custom-made systems that allow thousands of parallel laboratory experiments to be done in tiny tubes and vessels (see Gulati *et al.* 2009). Elegant experiments on oscillating devices in bacteria have shown how microfluidics can allow the behaviour of thousands of engineered cells to be monitored separately while also offering precise control of each cell's local environment.

Microfluidics is also expected to impact on the assembly of DNA parts. As well as BioBricks™ there are many other methods for assembling DNA into parts, devices and even up to the scale of genomes. The enzymes and chemical reagents required for DNA assembly are not cheap, and miniaturisation of reactions using microfluidics will cut costs and accelerate construction. Robotic automation of DNA assembly using liquid-handling robots or advanced microfluidics platforms is a major step to further increase the speed and decrease the cost of building prototype devices and systems. DNA assembly by a variety of methods has already been shown to be automatable using robotics, and CAD software can be used to link the design of constructs to programs that direct robot stations to, in turn, perform the DNA assembly from synthesised or shared standard parts. This will be aided by new standardised methods for DNA assembly that are amenable to both robotics and microfluidics and are suitable for working with all types of parts regardless of size, sequence or function (Fig. 3.9). Work published in 2010 by the J. Craig Venter Institute has described an automatable, sequence-independent DNA assembly method that can construct at every scale, going from custom DNA oligonucleotides to genes and all the

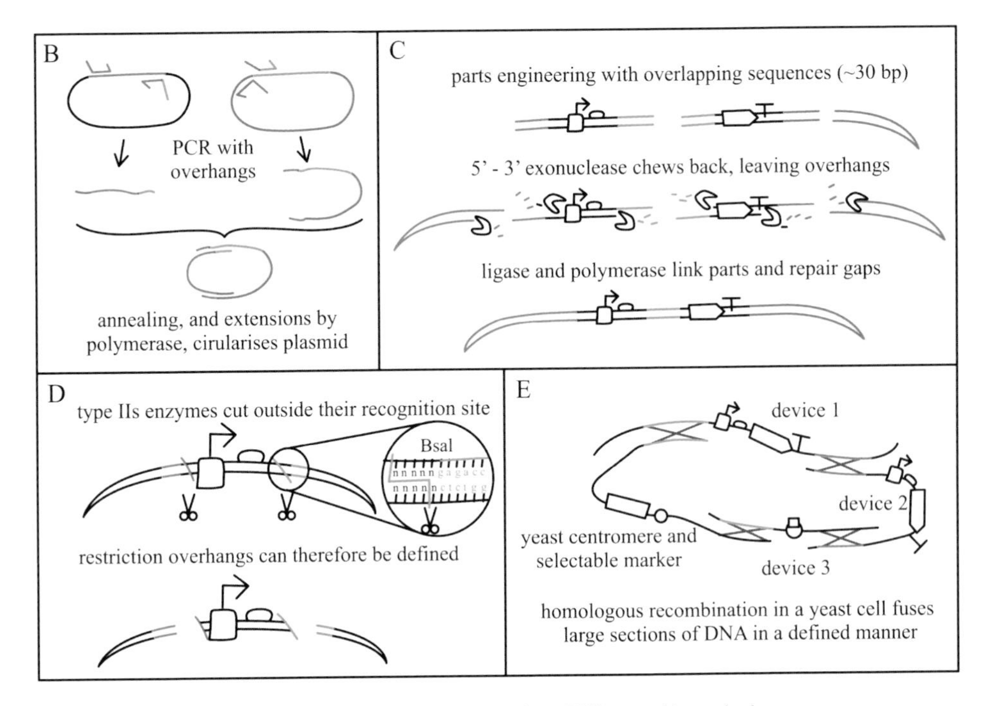

Fig. 3.9 Schematic diagram of various DNA assembly methods.

way up to a complete synthetic chromosome. New assembly methods such as this increase the capabilities of synthetic biology, opening up the possibilities of building constructs on a much larger scale.

A new branch of research, genome engineering, has already extended the capabilities of DNA assembly up to the scale of producing synthetic genomes from gene-sized parts. This has allowed synthetic biology researchers at the J. Craig Venter Institute to go as far as creating a living cell containing a genome entirely produced from custom-synthesised DNA. This landmark achievement is notable for synthetic biology as it opens up the possibility of producing customised cells completely from digital designs. Often called synthetic life, this idea of 'designer' cells with constructed genomes is attractive for engineering biology as researchers will have far greater control and understanding of their chassis cell if they have built it up from first principles. In theory this will allow new parts and devices to be added to a cell with greater predictability, especially if this can be combined with multi-scale, whole-cell models. Although our current understanding of some cells is quite extensive, especially for yeast and *E. coli* bacteria, we still rarely know enough to predict accurately how synthetic biology designs will perform in a cell in different environments. Designer organisms and minimal cells are a logical step to improving predictability. Removing all the unessential natural genes that could interfere with synthetic devices will make design and realisation of synthetic biology easier. This would also be advantageous for the safety of synthetic biology, as any genes that could be toxic or potentially cause disease could be removed and the synthetic cells could even be built to be orthogonal to natural systems. For synthetic biology, the notion of orthogonality is to design important processes in devices, systems or even whole synthetic cells so that they no longer interface with natural processes. Researchers have described modified ribosomes that translate synthetically designed mRNAs but do not work with natural ones, and further extensions of this could include using non-natural DNA and RNA bases for encoding synthetic biology designs separately to natural systems. Orthogonality is particularly useful for preventing crosstalk with natural systems, and for synthetic biology this would both enhance predictability of designs while alleviating safety fears of engineered biology contaminating natural life.

Reading

BIOFAB initiative. An open repository for professionally characterised biological parts. Available at: http://biofab.org/ [Accessed 18 August 2011].

Carlson, R. (2010). *Biology is Technology: The Promise, Peril, and New Business of Engineering Life*. Harvard University Press, Cambridge, MA.

Ellis, T, Adie, T and Baldwin, GS. (2011). DNA assembly for synthetic biology: from parts to pathways and beyond. *Integrative Biology* 3: 109–118.

Endy, D. (2005). Foundations for engineering biology. *Nature* 438: 449–453.

Gulati, S, Rouilly, V, Xize, N, *et al.* (2009). Opportunities for microfluidic technologies in synthetic biology. *J R Soc Interface* 6: S493–S506.

Ideker, T. (2004). Systems biology 101 — What you need to know. *Nat Biotechnol* 22: 473–475.

OpenWetWare. A wiki-style online community for bioscience. Available at: http://openwetware.org [Accessed 18 August 2011].

Registry of Standard Biological Parts. Available at: http://partsregistry.org/Main_Page [Accessed 18 August 2011].

Saccharomyces Genome Database (SGD). An open online database hosting the yeast genome sequence and associated information. Available at: http://www.yeastgenome.org/ [Accessed 18 August 2011].

CHAPTER 4

Minimal Cells and Synthetic Life

4.1 Introduction

In this chapter we will discuss the concepts of synthetic life and minimal cells. Some of the more controversial and thought-provoking topics of synthetic biology and modern science fall into this category, including the creation of synthetic cells and the search for the minimal biological system that can be called life. Ultimately, bottom-up synthetic biology aims to give us new cells that are rationally engineered to act as specialised chassis. These would be more useful to biotechnology than existing natural cells for a variety of reasons. A few examples of the advantages of synthetic cells are:

(i) Using engineered cells would be more predictable than using natural cells that we don't fully understand.
(ii) Synthetic cells can be engineered to be streamlined for one task only, allowing them to use fewer resources than natural cells.
(iii) Cells constructed from the bottom up can be made dependent on specific conditions, which can be used to prevent them from thriving outside of the desired environment.

Bottom-up synthetic biology broadly falls into two categories: (i) synthetic cells, which are newly designed microbes containing entirely synthetic DNA; and (ii) *in vitro* life, where biochemical reactions placed together can carry out the functions of life and act as artificial cells. To understand both of these it is also worth first considering the question: What is the minimum that we call life?

4.2 Minimal Cells

When we think of life there are millions of examples that jump into our heads, from massive multicellular organisms like elephants, trees and whales down to single-celled ones like bacteria, algae and yeasts. All of these share commonality; they contain a DNA program, they have RNA and protein tools and they are surrounded by a lipid membrane to form a cell. Smaller than all cells are viruses, which also typically contain either DNA or RNA and protein. Viruses act by infecting and reprogramming cells, but cannot reproduce without them. So viruses, along with transposons and prions, are not classed as free-living, and so typically they don't count as life, instead being something more like aberrant genetic programs.

The idea that all free-living life is based on cells is backed up by strong evidence that everything natural that we consider living contains at least one cell. Compartmentalisation of DNA, RNA and protein appears to be a fundamental property of life and is one of the key requirements for evolution. But how many parts are needed to make a cell, and what is the minimum possible cell that acts as free-living life?

4.2.1 *Natural minimal cells*

The exemplar chassis (host organisms) used in synthetic biology and biotechnology are *E. coli* bacteria and *S. cerevisiae* yeast. Both are single-celled microbes but are quite different in size. *E. coli*, a prokaryote, is about 1 micrometre in diameter and has a genome containing 4.6 million base pairs of DNA, encoding approximately 4,500 genes. Although one of the simplest examples of a eukaryote, *S. cerevisiae* is considerably larger and more complicated than most bacteria. It is about 10 micrometres in diameter, with a 12.5 million base pair DNA genome encoding approximately 5,800 genes. What is interesting to see is that even though yeast has 3 times more DNA and between 100 to 1,000 times more volume than *E. coli*, it only has around 1,000 more genes. Does this mean there is a fundamental number of genes we expect every microbe to have? To answer this, researchers have explored the diversity of microbes in search of free-living cells with much less DNA than *E. coli* and yeast. By studying these natural examples and counting the number of genes encoded on their genomes they have begun to discover the minimal limits of life, that is, just how many genes are required for natural function of free-living cells.

Minimal eukaryotic cells have genomes of around 12 million base pairs of DNA and contain about 5,000 genes. *S. cerevisiae* yeast is one of these, although a close relative, *S. pombe* yeast, has a slightly smaller genome. A few parasitic eukaryotes have considerably smaller genomes with half as many genes but, like viruses, these are dependent on their hosts, having done away with essential genes required for a cell to be free-living. An interesting example of a minimal eukaryote is *Ostreococcus tauri*, the smallest known free-living photosynthetic eukaryote (Fig. 4.1). This primitive alga has cells that are as small as *E. coli*, yet each contains the chloroplast and starch granule organelles seen in much larger plant cells. The genome of *O. tauri* is 12.5 million base pairs of DNA but this contains as many as 8,000 genes. *Ostreococcus tauri* is an example of a physically minimal cell that actually has quite a high number of genes.

Prokaryotic cells with the 'smallest' genome and the 'fewest' genes seem to be discovered every few years. It seems that the more we search through nature, the smaller the cell we can find. Two recent examples discovered are *Hodgkinia cicadicola* with 188 genes and a genome of 144,000 bp and *Carsonella ruddii* with 213 genes and a genome of 144,000 bp. Both of these microbes show just how few genes are required to encode the workings of a cell; however, both are not free-living but have entered into parasitic or symbiotic relationships with other cells — effectively being dependent on genes encoded elsewhere. A thermophilic archea prokaryote, *Nanoarchaeum equitans*, is also an obligatory

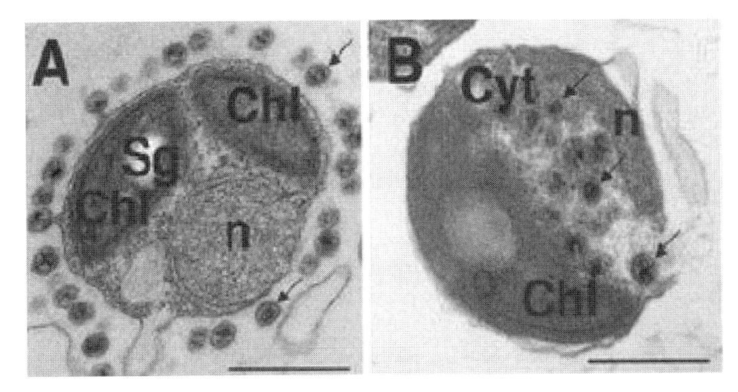

Fig. 4.1 Electron microscope images of *Ostreococcus tauri* cells. Organelles are labelled as Chl (chloroplast), Sg (starch grain), n (nucleus) and Cyt (cytoplasm). (Courtesy: E. Derelle *et al.* 2008. *PLoS ONE* 3(5):e2250.)

symbiont but is worth mentioning as it is the smallest known living organism, with cells only 400 nanometres in diameter. Its genome is only 490,000 bp long, containing just over 500 genes. Perhaps the smallest known microbe that is truly free-living is *Pelagibacter ubique*, which is a bacterium that lives in the sea and fresh water worldwide. It has a genome of around 1.3 million base pairs with about 1,300 genes. The cells are very small, less than a micrometre in length and only 200 nanometres wide, but they thrive across our planet and *P. ubique* is thought to be the most abundant organism on Earth, even accounting for around 50% of all cells in the temperate oceans in summer.

The natural minimal cells that we understand in the most detail are the *Mycoplasma* genus of bacteria, often responsible for human illnesses. *Mycoplasmas* are usually parasitic bacteria, so are strictly not classed as free-living. However, they are a good example of minimal cells as they can easily grow independently in laboratories when given a source of nutrients. In this genus, the species that have interested researchers looking at minimal life are *M. pneumoniae*, *M. mycoides*, *M. capricolum* and *M. genitalium*. *M. mycoides* and *M. capricolum* have properties that make them suitable for the laboratory experiments required to produce the first cells with a synthetic genome (see Section 4.3.1 of this chapter). *M. pneumonia* has a genome of only 816,000 DNA base pairs, encoding approximately 680 genes and has been the focus of in-depth studies of the workings of a minimal cell. It was one of the first organisms to have its genome sequenced and has also been the model organism for extensive systems biology investigations aimed at understanding the complete workings of a cell (Fig. 4.2). Lastly, of all the *Mycoplasmas*, *M. genitalium* has the smallest genome, and it is one of the smallest of any bacterial cell and one of the very first genomes sequenced. The genome has 583,000 base pairs of DNA encoding only 482 genes and it has become the focus of the minimal genome project.

4.2.2 *Genome reduction*

An active area of research is to determine the minimal number of genes required in a living cell or to find the minimal genome (either for all life or for a specific organism). One such project has been underway since the 1990s, led by Nobel Laureate Hamilton Smith. It focuses

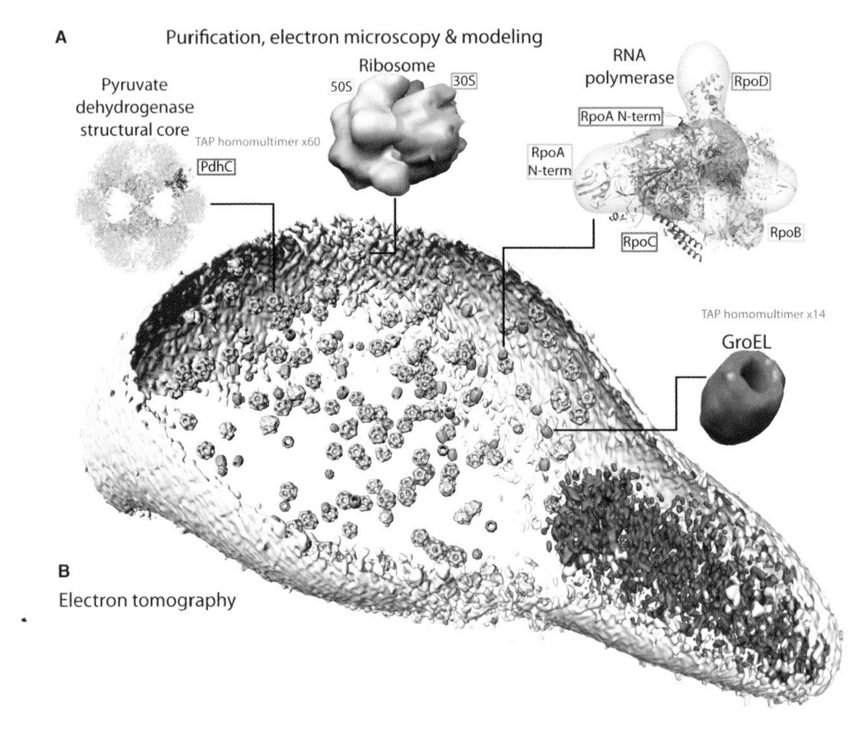

Fig. 4.2 *Mycoplasma pneumonia* cell structure (Courtesy: *Science*, 27 November 2009, 326(5957): 1235–1240. Reproduced with permission.)

on taking natural cells with small genomes and reducing the number of genes in these further until only the essential genes for life remain. Smith's project has centred on *M. genitalium* and its 482 genes. *M. genitalium* contains the necessary genes for the central dogma of life — DNA replication, transcription and translation. It also contains the genes for metabolising phospholipids, utilising vitamins and for undergoing glycolysis. But by being a parasite it has managed to remove many of the genes needed for amino acid and nucleic acid biosynthesis and it does not have the enzymes required for oxidative metabolism. It also has very few regulatory genes like transcription factors or two-component systems.

To determine which of the 482 genes are essential for minimal cellular life, researchers can take two approaches: a comparative approach or a reductionist approach. In the comparative approach, the genes and genomes cells are compared using DNA sequence data obtained from genome sequencing. Comparative analysis of all sequenced genomes reveals genes that are common to all life, such as DNA polymerases and the genes encoding ribosomal proteins and tRNAs, but it also highlights those genes whose functions are not currently known yet and are also clearly important. In particular for the *M. genitalium* minimal genome project, comparing the genomes of the closely related *Mycoplasma* species is very useful. A good indicator for genes that are essential are those that are found in *M. pneumonia* and have nearly identical DNA sequences to those in *M. genitalium*.

The lack of change in DNA sequence of genes is indirect evidence that they are important essential genes, as it shows that they have resisted mutational change through evolution. The comparative approach using all sequenced genomes surprisingly identifies only about 60 or 70 genes shared by all living organisms. This number is much lower than the minimal set needed, simply because different forms of life have evolved different genes to carry out similar functions. For example, the DNA polymerases in a human cell are completely different to those in bacteria. Focusing on *Mycoplasma*, which are very closely related, the comparative approach gives a more realistic figure of the minimal set of genes required — around 200 to 300.

Whilst the comparative approach uses genome sequence data to infer essential genes, the reductionist approach uses experiments. In the reductionist approach, researchers use molecular biology techniques to systematically remove individual genes in the genome to identify those essential for cell survival under lab conditions. To remove each gene, a gene 'knock-out' is performed. For the example *M. genitalium* minimal genome project, each knock-out was done by using a transposon that inserts a deleterious DNA sequence into the middle of each gene, and so destroys its function. For similar work done in other bacteria, such as in *Bacillus subtilis*, genes have been systematically knocked-out using a series of non-replicating plasmids designed to target each gene. Knock-out experiments in *B. subtilis* identified only 271 essential genes out of around 4,800, yet no one has attempted to remove all of the 4,500+ non-essential genes simultaneously and it is doubtful that the bacteria would still survive losing even half of these at the same time. The equivalent study in *M. genitalium* revealed that over 100 genes were not essential for the cell to survive, giving an approximate number of essential protein-coding genes of around 370. Interestingly, the function of about 100 of these 370 genes is still not really understood, indicating that our understanding of what is needed to make a minimal cell is still incomplete.

Whereas the deletion of all non-essential genes in *B. subtilis* has not been attempted, the aim of the minimal genome project in *M. genitalium* is to do exactly that and create an engineered minimal cell as a chassis for synthetic biology. The aim of the project is to create '*M. laboratorium*', a *mycoplasma* with as few genes as possible. While this could be achieved by sequential deletion of genes from the natural *M. genitalium* genome, the researchers involved have instead decided to build up this minimal genome from its parts to create a minimal synthetic genome.

4.3 Synthetic Life

The definition of synthetic life is contentious, although all versions acknowledge that to be synthetic requires some element of design and construction by humans. Is synthetic life whole cells completely assembled from chemicals, is it simple self-sustaining biochemical processes that we have put together or is it something else? To give a clear story in this chapter we refer to synthetic life as replicating cells that contain a DNA genome put together from synthesised parts. After all, cells are fundamental to life and their functions are encoded by their genome. A man-made genome is therefore the starting point for synthetic life, and the story of the first synthetic cellular genome is an excellent example that bottom-up synthetic biology is possible.

4.3.1 *Genome synthesis*

Researchers re-engineer the genomes of a variety of microbes and other cells on a daily basis in laboratories around the world. Swapping DNA parts in and out of existing genomes has been possible for decades, as has the ability to introduce large regions of new DNA into cells. At the same time that molecular biology has developed the techniques to rearrange the DNA inside cells, chemistry has become adept at synthesising new DNA sequences directly from single molecules (see Chapter 3). With the falling cost of DNA synthesis and the increasing improvements in molecular biology, it was inevitable that researchers would consider synthesising whole genomes of cells from single molecules.

The first synthetic genomes constructed were viral genomes, with the 7,500 base pair poliovirus genome synthesised in 2002, and the 5,400 base pair phi X174 phage genome synthesised in 2003. These DNA genomes were assembled in the lab from chemically synthesised DNA oligomers, using the same molecular biology technologies used in gene synthesis. Placing these DNA genomes into appropriate bacterial host cells (by transformation) leads to the production of new copies of functioning viruses, even though the initial DNA encoding the information was assembled in the lab by chemistry. Repeating this same feat, not for viruses but for cells, is a much larger task but one that has been successfully achieved by the same team involved with the minimal genome project.

To synthesise the first complete cellular genome, researchers based at the J. Craig Venter Institute (JCVI) again turned to *M. genitalium*. To synthesise its circular genome, its 583,000 base pairs of DNA were divided into 101 arbitrary regions, each about 6,000 base pairs long. Each region overlapped slightly with the next, meaning that they both contained a few hundred base pairs of the same DNA sequence. These 101 regions were synthesised chemically by several DNA synthesis companies and were checked by DNA sequencing. Using a cocktail of enzymes in the lab, these regions were assembled together using the overlapping sequence to define which region goes next to which. In rounds of DNA assembly, the 101 regions made 24 larger sections that were stepwise assembled into four quarter-genome pieces. These four pieces were then simultaneously inserted into yeast cells (co-transformation), which uses its own DNA repair enzymes to link overlapping DNA fragments together to form complete circular genomes. The chemically synthesised *M. genitalium* genome was designed to include a few extra DNA parts to allow it to be hosted in yeast cells, to be selectable using antibiotics and to replicate when yeast cells divide. This assembly and then extraction of this genome from yeast cells, revealed in 2008, resulted in the first chemically synthesised cell genome.

Unlike viral genomes, a chemically synthesised cellular genome cannot simply be added to growing bacteria to create synthetic life. The genome on its own is only a DNA instruction book and requires the enzymes and the environment of a cell to be read. In order to start synthetic life from a synthetic genome, an empty cell is required to 'boot up' the system. Creating these empty cells and getting a whole genome into them and working is a technical challenge in itself. It was first demonstrated in 2007, when the natural DNA genome from *M. mycoides* was taken from its own cell, cleaned and then used to transform *M. capricolum* cells. Following a complex transformation method, some *M. capricolum* cells were seen to change and resemble *M. mycoides* cells, and later these were verified to be indistinguishable from natural *M. mycoides*. This process of 'genome transplantation' was combined with the chemical genome synthesis method described above in 2010 to create the world's first synthetic life. The JCVI research team assembled the complete 1 million base pair *M. mycoides* genome from synthesised DNA fragments and then used this to transform *M. capricolum* into *M. mycoides*. As well as including a few extra DNA parts to allow the genome to be hosted in yeast and selectable by antibiotics, the researchers included coded messages in the DNA sequence to act as 'watermarks' confirming the cells did have a man-made genome (Fig. 4.3).

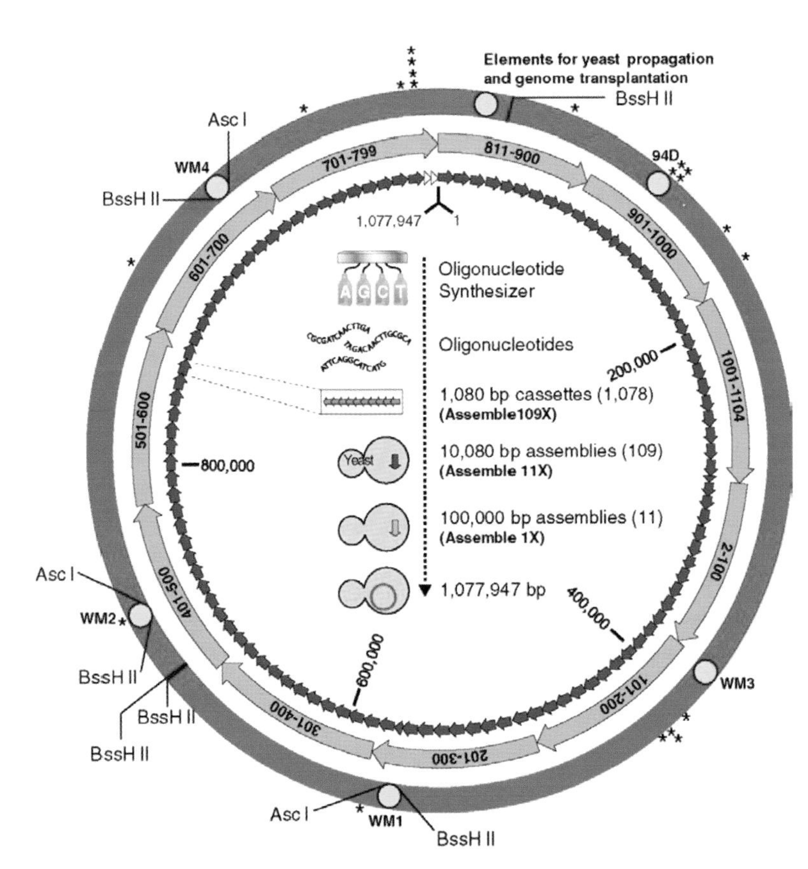

Fig. 4.3 Construction of a synthetic *M. mycoides* genome. The outer circle represents the 1 million base pair circular DNA genome. The genome has been edited in a number of places to include watermarks (WM). (Courtesy: *Science*, 2 July 2010, 329(5987): 52–56. Reproduced with permission.)

4.3.2 *Designer cells*

Synthetic life means more than just the chemical synthesis of existing bacterial genomes; it opens up the possibility of creating whole new organisms containing genomes never before seen in nature. While researchers have already shown that a synthetic genome containing a few extra DNA parts and watermark sequences can become the genome of a *Mycoplasma* cell, the next step for synthetic life is to produce cells with tailor-made genomes. An early target for such a 'designer' cell is the theoretical *M. laboratorium* from the minimal genome project described above. Synthesis of a *Mycoplasma* containing this reduced genome will give a new minimal cell and greatly improve our understanding of what is required for cellular life. If the synthesis and construction of the genome is done using a rational parts-based approach, it should be easy to remove genes further from this cell or to add new ones as needed. The cell and its genome would be an ideal chassis for synthetic biology as its reduced number of genes would mean less complexity and would be, therefore, more straightforward to mathematically model and simulate. To such a chassis it could be possible to predictably add many new genes, such as pathway genes that direct the resources of the cell into the production of fuel compounds. With the technology of complete genome synthesis, these new functions could be written into the genome of the cell and designed from the bottom up.

The rational engineering approach of synthetic biology and the ability to have customised cell chassis fit hand-in-hand. From these the logical future is designer cells, synthesised or modified to have the necessary attributes for specialised functions. While the early achievements in this field have been with *Mycoplasma* bacteria, the methods should be applicable to a range of bacteria and other microbes like yeast. One example of a designer cell is a soil bacteria that also contains the genes required for photosynthesis, and so could be used as a fuel-generating microbe. An attempt to make this novel strain of bacteria in 2005 was unsuccessful. Researchers tried to fuse the entire 3.5 million base pair genome of the photosynthetic *Synechocystis* cyanobacteria into the 4.2 million base pair genome of *B. subtilis* to create a hybrid with dual functionality. Unfortunately, most of the cyanobacteria genes were not expressed (silent) in the soil bacteria and so photosynthesis was not seen. Despite this early failure, the idea illustrates the potential of bottom-up synthetic biology to create new cells with desired functions.

As well as adding new functions to custom cells in a logical approach, bottom-up synthetic biology can give us chassis cells with other useful properties. Minimal cells are inherently streamlined and can grow with greater efficiency than those containing many more genes. For the production of fuels, drugs and other high value compounds using synthetic biology, it will be advantageous to have the chassis cell take up as little energy as possible for its own growth and maintenance. This allows the resources, such as sugars, to be maximally converted into product. Designer cells also offer the possibility of more stable engineering for synthetic biology. By constructing genomes with recombination sites and transposon sequences removed and added DNA repair enzymes, it should be possible to create a cell chassis more robust against evolution due to DNA mutating and moving around in cells. For synthetic biology applications like biosensors it is desirable that the genetic structure of the engineered cells remains stable over many generations, so specialised chassis cells that do not mutate would be a useful option. Contrary to this, it would also be desirable in other circumstances to have custom cells that had little protection against mutation. Chassis cells could be engineered specifically for fast evolution, containing many recombination hotspots and lacking DNA repair mechanisms. Directed evolution is already a key technology in drug discovery and metabolic engineering, using mutation and selection to create new genes that encode for specialised enzymes. This evolution could be programmed into designer cells to enable drug discovery to be optimised in cells by synthetic biology.

4.4 Origins of Life in Nature and in the Lab

4.4.1 *The RNA world*

Contemporary theories on biological origins suggest that life arose via a self-replicating molecule or assembly of self-replicating molecules. It has been suggested that nucleic acids or nucleic acid-like molecules were the first self-replicators because their inherent complementarity gives an automatic template from which to copy the next generation (although some researchers hold that self-replicating peptides or lipid systems may have preceded or emerged in parallel with nucleic acids). An early nucleic acid replicator is also tempting because of the prevalence of molecules in modern cells that are related to nucleic acids, such as ATP and other cofactors. Given the fact that the ribosome is at its core a ribozyme, it seems highly likely that there was a complex RNA world that preceded the modern world of protein catalysts. This RNA world may have in turn descended from an early nucleic acid replicator by duplication, parasitism and diversification.

Researchers have recapitulated the Darwinian evolution of RNA self-replicators as well as other functional RNAs. One of the earliest examples of extracellular evolution of molecules is the work done by the Spiegelman group. In this work, the researchers discovered that the bacteriophage Qβ utilised an RNA-dependent RNA polymerase to replicate its own RNA genome. In this simple system (containing replicase, template RNA and nucleotides) the researchers found that the system showed autocatalytic kinetics, indicating that self-propagation of the viral genome was occurring and could do so *in vitro* with the necessary components. Furthermore, the Qβ system was used to set up an extracellular evolution experiment where mutant templates could 'compete' with one another for the limited pool of replicases and nucleotides, eventually selecting for template sequences that were able to replicate faster than the original (parent) template.

4.4.2 *Chemical replicating systems*

The ability to evolve functional and catalytic molecules from libraries of synthesised variants has inspired efforts towards the development of chemical replicating systems. Several groups have shown that chemically synthesised lipid vesicles are able to grow by accumulation of additional lipid material and divide under gentle shear forces. This work demonstrates that what appears to be a very biological phenomenon (cellular compartment growth and division) can be achieved by simple chemical and physical systems. There are ongoing

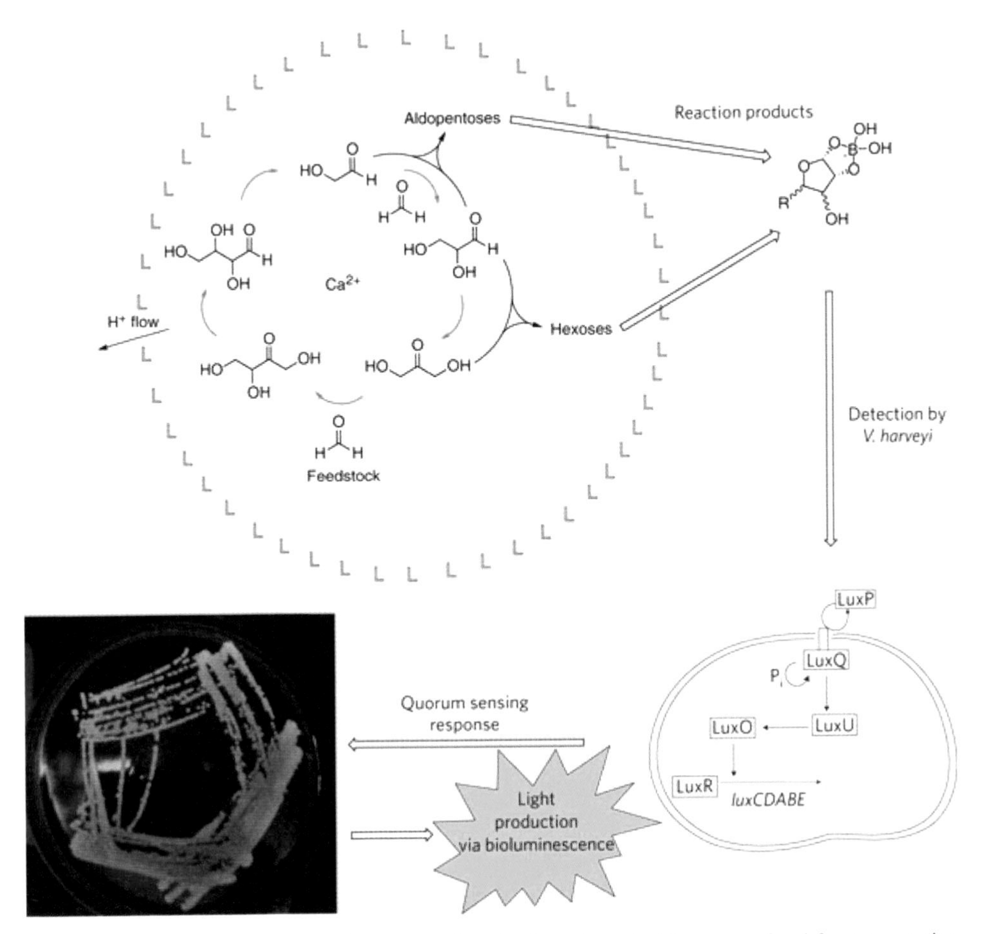

Fig. 4.4 Communication between a chemical 'cell' and a living bacteria. The encapsulated formose reaction produces carbohydrates able to trigger the *V. harveyi* quorum-sensing mechanism, which induces bioluminescence in the bacteria. (Courtesy: *Nature Chemistry* 1, 377–383, 2009. Reproduced with permission.)

efforts to include molecules capable of self-templating and replication in these vesicles. A likely candidate for a self-templating and self-replicating molecule is a ribozyme capable of polymerising RNA (see above) — in this case, a single ribozyme can act as the template and the catalyst for copying itself.

The development of chemical replicating systems will be useful for studying the minimal set of matter and information that can constitute self-replication and may shed insights on the origins of the first forms of life. There is also nascent work to engineer chemical replicating systems to interact with biological systems. Researchers have recently engineered a membrane-encapsulated chemical 'metabolism' that is able to synthesise complex carbohydrates from the feedstock formaldehyde, and trigger the quorum-sensing mechanism of the bacteria *Vibrio harveyi* (Fig. 4.4). The chemical cell was able to trigger quorum sensing and bioluminescence in bacteria, indicating that a completely artificial 'life-like' system was able to interact with a living system. The interaction between living and non-living systems will be an important area for synthetic biology in the future.

Reading

Forster, AC and Church, GM. (2006). Towards synthesis of a minimal cell. *Mol Syst Biol* 2(45). Available at: http://www.nature.com/msb/journal/v2/n1/pdf/msb4100090.pdf [Accessed 10 October 2011].

Gibson, DG, Glass, JI, Lartigue, C, *et al.* (2010). Creation of a bacterial cell controlled by a chemically synthesized genome. *Science* 329(5987): 52–56.

Glass, JI, Assad-Garcia, N, Alperovich, N, *et al.* (2006). Essential genes of a minimal bacterium. *Proc Natl Acad Sci U S A* 103(2): 425–430.

Jewett, MC and Forster, AC. (2010). Update on designing and building minimal cells. *Curr Opin Biotechnol* 21(5): 697–703.

Kuhner, S, van Noort, V, Betts, MJ, *et al.* (2009). Proteome organization in a genome-reduced bacterium. *Science* 326(5957): 1235–1240.

Moya, A, Gil, R, Latorre, A, *et al.* (2009). Toward minimal bacterial cells: evolution vs. design. *FEMS Microbiol Rev* 33(1): 225–235.

Ochman, H and Raghavan, R. (2009). Excavating the functional landscape of bacterial cells. *Science* 326(5957): 1200–1201.

CHAPTER 5

Parts, Devices and Systems

5.1 Introduction

Synthetic biology follows a hierarchical structure, building up systems from smaller components. At the lowest level are the parts, which are pieces of DNA that encode for a single biological function such as an enzyme or promoter. These parts are then combined into the next layer, a device, which is a collection of parts that performs a desired higher order function, for example production of a protein. Devices are further combined into a system, which can be defined as the minimum number of devices necessary to perform the behaviour specified in the design phase. Systems can have fairly simple behaviour (e.g. an oscillator) or more complicated behaviour (e.g. a set of metabolic pathways to synthesise a product). Parts and devices are usually treated as modular entities in design and modelling. This means that it is assumed that they can be exchanged without affecting the behaviour of other system components that are left untouched, which is problematic in biological systems. Systems must be implemented in a chassis, which provides the underlying biology necessary to transcribe and translate the system as well as any enzymatic substrates that would be necessary. The chassis can be a living organism (also called *in vivo* implementation), or it can be abiotic, providing only the necessary biochemical components for *in vitro* transcription and translation.

5.2 Parts

A part is a piece of DNA that encodes a biological function. In synthetic biology, parts are usually not used singly, but instead are combined into devices. When following the design cycle outlined in Chapter 2, the part is the unit which undergoes extensive characterisation and testing in different conditions to make a datasheet in order to enable modelling and model-based design. Some commonly encountered parts, along with their functions, are listed in Table 5.1 and Fig. 5.1 (which shows the symbols used to denote each part in the MIT Registry of Standard Parts created from the International Genetically Engineered Machine (iGEM) competition).

Table 5.1 Some common biological parts.

Part	Function	BioBricks™ Example
Promoter	Initiate transcription	BBa_J23101 *E.coli* constitutive
Operator	Regulate transcription	BBa_J64987 LacI binding site
Ribosome binding site	Initiate translation	BBa_J63003 S cerevisiae designed
Protein coding sequence	Encode for a protein	BBa_E0040 GFPmut3b
Terminator	Attenuate transcription	BBa_B1001 *E.coli*, artificial, small

Parts

Promoter Ribosome binding site Protein coding sequence Terminator

Example Devices

Protein generator

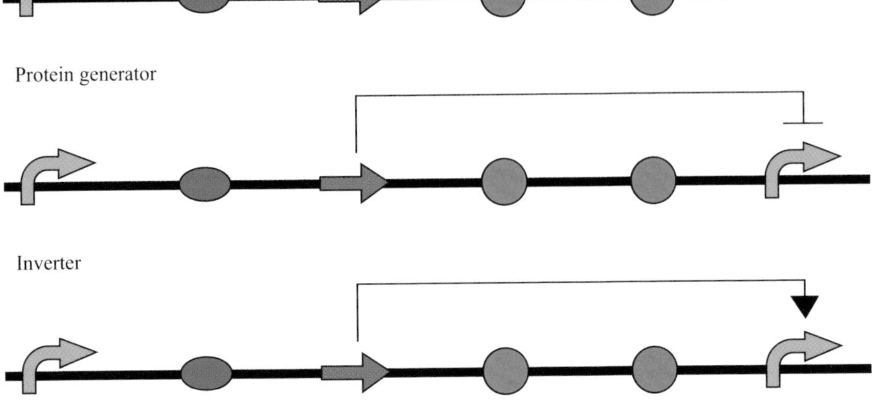

Inverter

Receiver

Fig. 5.1 iGEM symbols for parts and some example devices.

5.2.1 *Promoter*

A promoter is a sequence of DNA that recruits RNA polymerase and other accessory proteins to prime the transcription of messenger RNA (the molecular details of this process are described in Chapter 1). There are two common types of promoters — regulated and constitutive. Constitutive promoters are not regulated by any DNA binding proteins and, therefore, are always 'on'. Constitutive promoters are useful in synthetic biology for providing a constant level of a protein of interest to a system. By varying the sequence of the DNA encoding the RNA polymerase binding site, it is possible to vary the strength of the promoter and to increase or decrease the number of mRNA copies that are made from it per second. This, in turn, will lead to a different steady-state concentration of protein after the effects of translation and degradation are applied.

Regulated promoters often contain additional DNA sequences, known as operators (see below), which allow other protein factors to bind to the DNA and modulate the transcription rate. Thus, the timing of transcription can be controlled, either through the addition of a small molecule, which affects the recruitment of accessory factors to the operator, or by using a second genetic circuit to produce the accessory proteins at specific times. Regulated promoters can be either positively regulated (factors binding to the DNA increase transcription rate) or negatively regulated (factors binding to the DNA decrease transcription rate). In many cases negative regulators are present within the cell and will prevent transcription from the promoter under normal conditions. In this case, the default state of the promoter is 'off'. However, perturbing the biochemistry of the chassis (by adding small molecules or changing the temperature, pH, etc. of the system) can sometimes affect the interaction of the regulator with the operator, turning the promoter to the 'on' state. In this case, the promoter would be termed inducible.

5.2.2 *Operator*

Classically, a DNA sequence is termed an operator if it recruits the binding of factors that negatively regulate the promoter, also known as repressors. However, DNA sequences that bind positive regulators, also called activators, are sometimes also referred to as operators. Often in synthetic biology, the operator will not be considered as a separate part, but rather as a part of the promoter and the promoter/operator set will be characterised together as a single unit. The strength of repressor or activator binding is influenced by the DNA sequence of the operator. This can be manipulated to tune the level of repression or activation and, thus, the strength of the promoter.

5.2.3 *Ribosome binding site*

A ribosome binding site is a sequence of mRNA that recruits ribosomes for the initiation of protein translation. In prokaryotes, the ribosome binding site is also called the Shine–Dalgarno sequence (named after Dr John Shine and Dr Linda Dalgarno who discovered the sequence in 1975) and in eukaryotes it is known as the Kozak consensus sequence (after discoverer Dr Marilyn Kozak). Like promoters and operators, the sequence of the ribosome binding site affects the strength of ribosome binding and, therefore, the amount of protein synthesised.

5.2.4 *Protein coding sequence*

A protein coding sequence is a sequence of DNA that encodes for a protein. It begins with a start codon (ATG) to initiate protein translation, and ends with one of three stop codons (TAA, TAG or TGA) to indicate the end of the translation region. In between are the DNA triplets (codons) which denote the amino acids that make up the protein. Three types of protein coding sequences are commonly used in synthetic biology: reporters, repressors and enzymes.

Reporters are used as readouts of dynamic biological events and usually consist of a protein that produces a visual or fluorescent output. The green fluorescent protein (GFP) and its derivatives are often used as reporters because they gain fluorescence spontaneously upon protein folding and do not require the addition of any exogenous reagents such as enzymatic substrates. Other reporters include enzymes which catalyse a reaction leading to a coloured or fluorescent product. One classic example is the enzyme β-galactosidase (also known as LacZ) which is the enzyme utilised in blue–white screening schemes for cloning. LacZ naturally catalyses the degradation of lactose into glucose plus galactose, but also hydrolyses galactose residues from a variety of chemical substrates. Several chemical analogues have been designed such that upon cleavage of the galactose residue a coloured or fluorescent product is released which serves as an indicator and allows for the calculation of the amount of LacZ present in the sample.

Repressors are very often used as components of synthetic devices such as logic gates, switches and oscillators (discussed below). Their natural function is to bind to the operator region of promoters and prevent (repress) transcription. In addition to their DNA binding sites, repressors also contain binding sites for small molecules (inducers) that can influence their ability to bind DNA. The small molecules can be used to control the action of the repressor, switching it on and off. The final type of coding region used in synthetic biology is an enzyme, which performs an interesting chemical reaction for building systems. For example, a set of enzymes that perform sequential reactions form a metabolic pathway.

5.2.5 *Terminators*

A terminator is a sequence of DNA that signals the end of a transcriptional unit by causing the RNA polymerase to cease transcription and disengage. Terminators are placed past the 3' end of a protein coding sequence. In prokaryotes, terminators are often palindromic in DNA sequence, which leads to the formation of strong secondary structure (e.g. hairpins) that help to weaken the interaction between the RNA polymerase and the DNA strand being copied. Often devices will incorporate two tandem terminators to reduce the chance of transcriptional read through.

5.3 Devices

Parts are assembled into devices which perform simple, user-defined functions. Figure 5.1 depicts a schematic of a few common devices using the same symbols to describe common DNA parts. Note that devices can have similar architecture but perform different functions, depending on the function of the individual parts. For example, the schematics for an inverter and a receiver are largely the same. However, their functions are different. In this case, the difference in function comes from the fact that the protein generated in the first half of the device has a different interaction with the promoter in the second half of the device. In the case of the inverter it represses the promoter, whereas in the case of the receiver it binds and helps initiate transcription from it.

5.3.1 *Reporters and protein generators*

The simplest device is a protein generator which combines a promoter, ribosome binding site, protein coding region and terminator. As a unit these function to produce a protein via transcription (promoter, terminator) and then translation (ribosome binding site, protein coding region).

5.3.2 *Logic gates*

The term 'logic gate' comes from electronic circuits and is used to denote a device where multiple inputs are assessed together in order to control a single output. These simple circuits can be combined to create more complex electrical circuits. They also form the basis of modern computers, which contain hundreds of thousands of logic gates. The operations performed by logic gates can be described with Boolean algebra using binary states: 1 if a condition is satisfied, 0 if it is not. A logic gate takes multiple binary inputs, assesses them according to a predefined rule and produces a single binary output: 1 for on, 0 for off. The identity of the predefined rule determines the type of logic gate. In synthetic biology, logic gates have been constructed from genetic circuits, biochemical networks and free nucleic acids.

5.3.2.1 *AND*

In an AND gate all conditions must be satisfied in order for the output to be switched on. The simplest AND gate is controlled by two inputs, both of which must be present in order to produce the output. The logic applied to this gate can be formulated as 'If A AND B then "ON", Else "OFF"'. In synthetic biology there have been a wide variety of AND gate designs. For example, an AND gate can be constructed as a hybrid promoter in which two different operator sites are used to control transcription (Fig. 5.2). Two different repressors can then bind to these sites and block transcription. The presence of both inducers is required to relieve repression, making the output dependent on the presence of two small molecules.

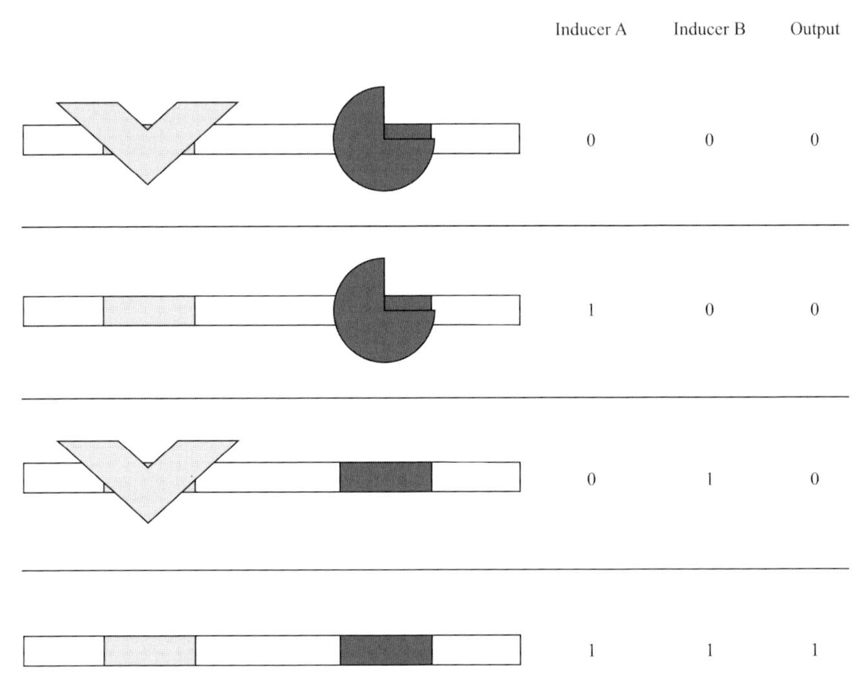

Fig. 5.2 AND gate promoter constructed from two operators (grey and black), which are bound by two repressors. The promoter remains off unless both inducers are present.

5.3.2.2 *OR*

OR gates follow the logic rule, 'If A OR B, then "ON", Else, "OFF"'. Therefore, for an OR gate to be switched on, only one of the input conditions must be satisfied. A very rudimentary OR gate can be constructed from a parallel set of reporter genes, each controlled by a single inducible promoter (Fig. 5.3). The presence of either molecule will turn the circuit on. Many other OR gate designs are possible.

5.3.2.3 *NOT (Inverter)*

A NOT gate takes a single input and inverts it. Therefore, it is also known as an inverter. When a signal is present, then the output of the NOT gate is 'off', whereas if the signal is absent, then the output is 'on'. NOT gates in synthetic biology have been constructed from compatible promoter/repressor pairs, like the example in Fig. 5.4. An inducible promoter that responds to the desired input signal controls the production of a repressor which controls the production of the output signal. The input signal causes the production of the repressor, which in turn represses the promoter that controls the output signal, turning it off. In the absence of the signal, the output signal remains on.

5.3.2.4 *Negated gates*

Logic gates are modular and can be combined to produce more complex behaviours. Connecting a NOT gate to any other gate will invert the function of that gate. A

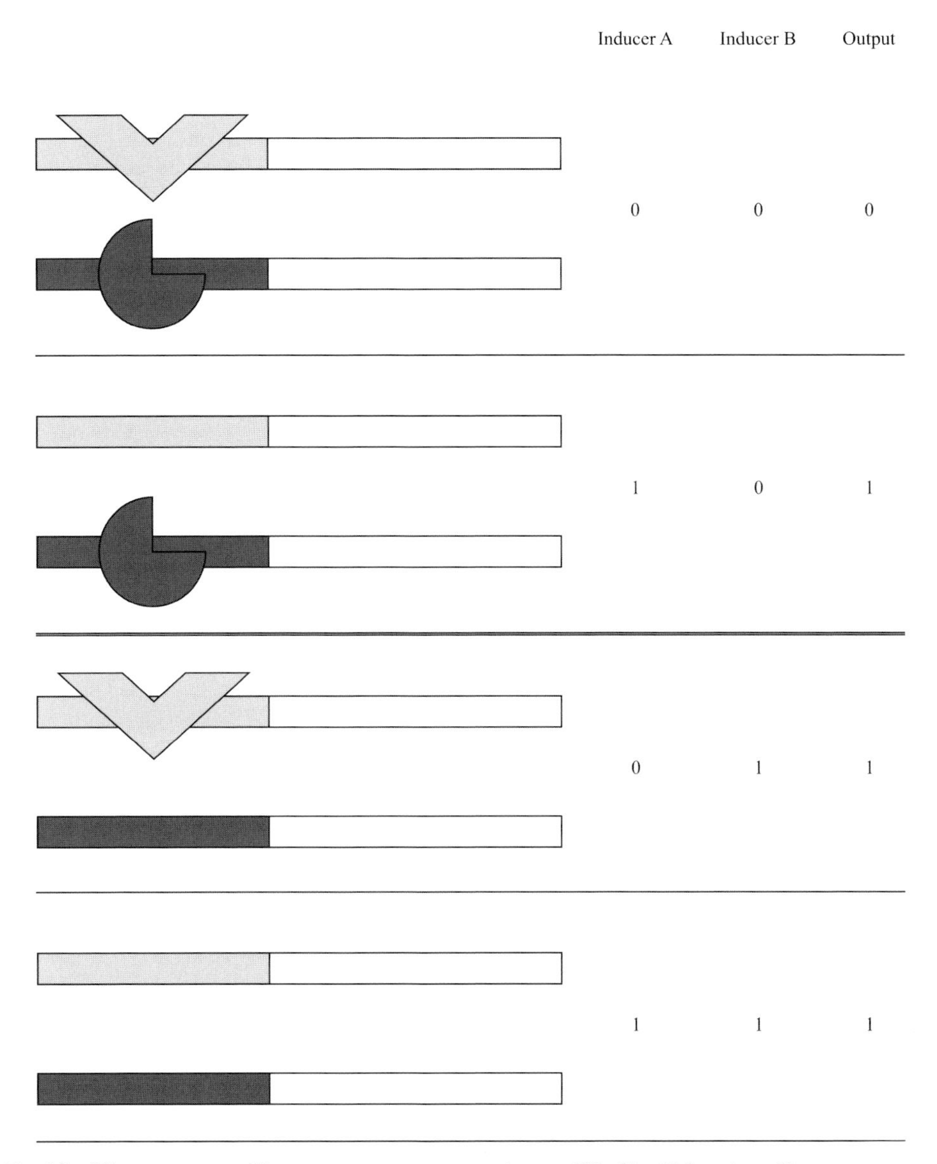

Fig. 5.3 OR gate constructed from two separate promoters (grey and black), which are bound by two repressors. The presence of either inducer is enough to switch the circuit on by allowing transcription from one of the promoters.

NOT gate can be combined with an AND gate to produce a NAND gate, which has behaviour the opposite of the AND gate depicted in Fig. 5.2. In other words, the NAND gate produces an output unless two signals are present. Similarly, a NOT gate can be combined with an OR gate to produce a NOR gate, whose behaviour is opposite of the OR gate in Fig. 5.3. A NOR gate will only produce a signal when both signals are absent.

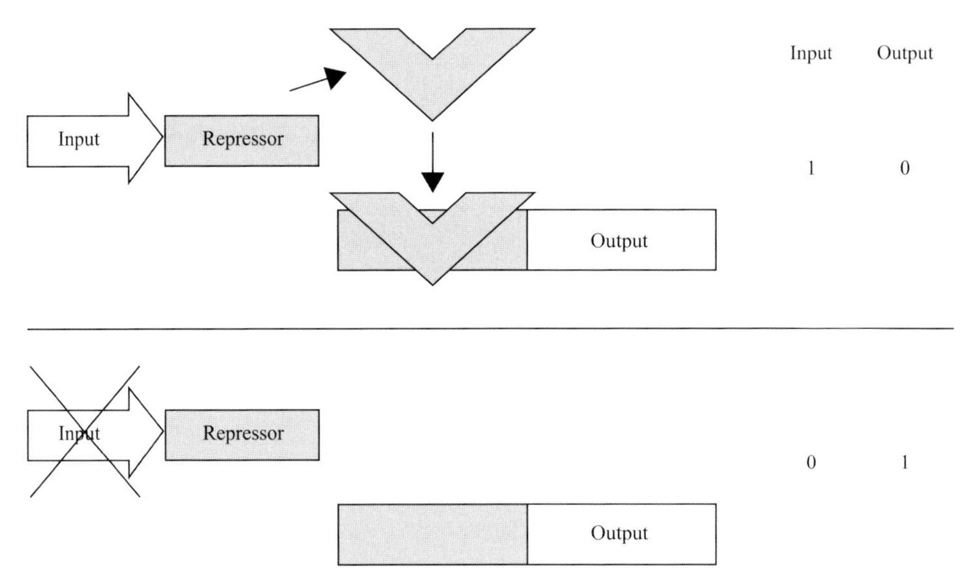

Fig. 5.4 A NOT gate based on a compatible promoter/repressor (grey) whose production is controlled by an input signal. The output is inverted, meaning it is absent when the signal is present and present when the signal is absent.

5.3.3 *Cell signalling senders and receivers*

Cell–cell signalling devices are useful in synthetic biology for coordinating the behaviour of populations. Most of these devices have been implemented in bacteria and rely on a bacterial signalling mechanism known as quorum sensing. Quorum sensing is used by cells to sense the population density and coordinate gene expression and other behaviours. Quorum sensing is accomplished via the synthesis, secretion and uptake of small molecules, the concentration of which affects the behaviour of the cell population. Thus, two types of device can be created from quorum sensing for cell–cell communication: senders, those that synthesise and secrete the molecule; and receivers, those that take up the signal and produce an output in response.

Natural quorum sensing is built around small organic molecules called acyl-homoserine lactones (AHLs). Different species of bacteria synthesise different AHL molecules that vary in the length of the acyl chain or in the chemical functional groups it contains. Thus, it is possible to create sender devices that synthesise and secrete AHL molecules that are not normally secreted by a given chassis, leading to an orthogonal system for behaviour coordination. Placing this device under the control of a regulated promoter allows the system to respond to environmental signals.

Receiver devices are based on transcription factors that bind the AHL signal molecules and turn on target gene expression in response. In natural quorum sensing these transcription factors are continually produced, but are unstable in the absence of the AHL molecule which they bind. When the AHL is present, they are stabilised, bind to the operator region of a target promoter and turn on transcription. Synthetic receiver devices consist of the promoter, which is controlled by the AHL responsive transcription factor upstream of genes for the desired response. Orthogonal receiver devices can be constructed by using transcription factors and promoters from quorum-sensing signalling pathways from other species. These are supplied as an additional part of the genetic circuit either under the control of a constitutive promoter or under the control of another inducible one.

5.3.4 *Light control devices*

Building complex systems requires control over the timing of gene expression. This can be achieved with some of the parts and devices described above. Inducible promoters (see Section 5.2.1) can be activated through the addition of small molecules to the growth medium, and receiver devices (see 5.3.3) can be activated through manipulating the sender devices. However, light offers a means of controlling expression that is non-invasive and potentially has a higher degree of spatial control. Thus, the development of devices to manipulate chassis in response to light has received a lot of attention with devices implemented in many of the commonly used chassis including bacteria, yeast and mammalian cells.

In basic terms a device for light control consists of a light-gathering element and an effector that controls transcription in response to the light input. Light-gathering elements to date have mostly been photoreceptors from organisms that naturally respond to light, usually plants or cyanobacteria, heterologously expressed in the chassis in question. These usually contain a chemical chromophore element that harvests the light itself, contained within a protein receptor. Often the absorption of photons by the chromophore results in a conformational change in the protein receptor, which helps to signal to the effector that light has been received.

Once the effector has received a signal from the light-gathering element, it then controls gene expression in response. To do this, the effector must interact with the native genetic machinery of the chassis, so in most cases natural elements from the host have been used and fused to the protein receptors to form chimeras. Two strategies, phosphorylation of transcription factors and dimerisation of proteins, have been used for light control to date, but in principle any strategy can be developed.

5.4 Simple Systems

A number of simple systems have been developed that are composed of one or more devices. These are discussed below. More complex systems, i.e. those that synthesise a product or vastly alter cellular behaviour, are discussed in Chapter 8.

5.4.1 *Feedback loops*

Feedback occurs when the output from a device influences the future behaviour of that device. These effects can be either negative (reducing the future output) or positive (increasing the future output). Feedback systems are inherent to biological organisms and theoretical analysis shows that they are necessary to reduce the noise or biological variability within a cellular system. This is supported by a wealth of evidence from control engineering where feedback loops are often used to increase the robustness of a system.

Many different designs for feedback loops can be developed depending on the overall desired behaviour of the system. Some naturally occurring proteins are generically useful for certain types of feedback. For example, repressors are useful for developing negative feedback loops by influencing the promoter used to produce them. In this case, the repressor is placed in an operon with the circuit of interest and its production will decrease the future level of transcription and translation of the circuit of interest. Similarly, activators can be used to produce a positive feedback loop when placed under the control of a promoter that they influence. Another scheme for positive feedback places specific RNA polymerases on the operon, which will increase the transcription of the circuit when produced.

5.4.2 *Switches*

Continuing the analogy with electrical engineering, a switch is a device that allows the cells to be altered from one state to another, i.e. turned on or off, usually in response to a user input or an environmental signal. A toggle switch is a switch with two arms arranged in feedback such that each controls the other. Depending on the conditions in the cell or the presence of an external signal, the build up of one regulator can 'overpower' the other, flipping the switch and leading to downstream effects. A schematic of one possible organisation of a genetic toggle switch is illustrated in Fig. 5.5. In this case, depending on whether signal A or

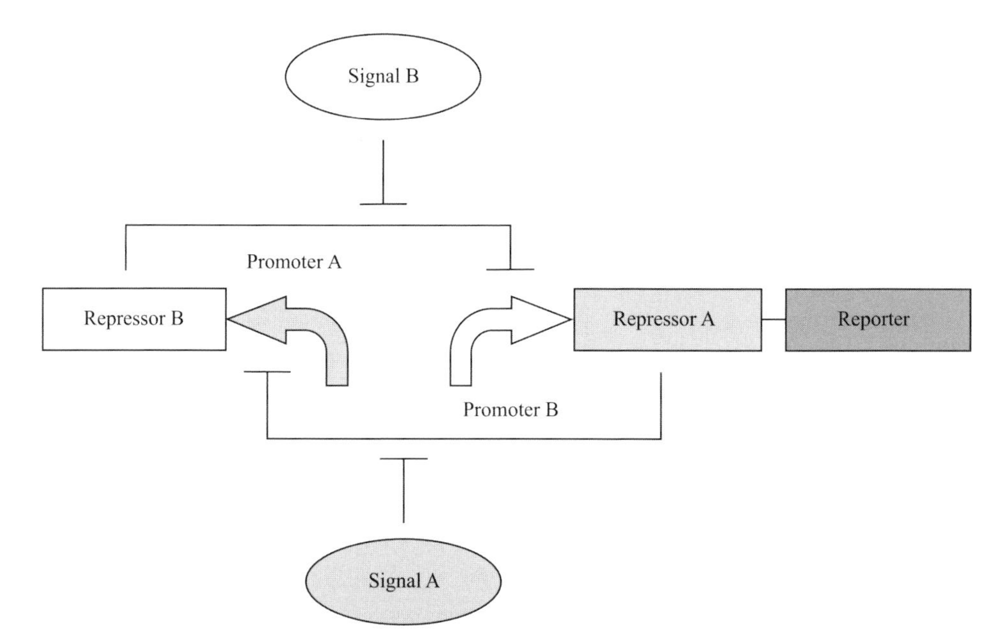

Fig. 5.5 A generalised toggle switch composed of two repressors, A and B, each of which interacts with a promoter, A and B, to repress transcription. This repression can be altered by a signal. A reporter can be placed in tandem with either or both repressors to give visual output of the switch function.

signal B is present and the strengths of promoters A and B, either repressor can accumulate, flipping the switch. A reporter can be placed in frame with either (or both) repressor to give a visual indication of when the switch has flipped. Theoretical studies have also shown that any network with an even number of repressors in feedback will display switch-like behaviour, so more complex switch designs are possible.

Toggle switches with delayed action can be used to trigger gene expression after a user-defined period of time. These are also called timer switches. The length of time before switching is controlled by the relative strengths of the two promoters, which lead to a differential accumulation of one of the repressors. This leads to the switching mechanism being activated when the concentration of repressor surpasses a threshold.

5.4.3 *Oscillators*

Oscillations are variations in the level of system output over time in a defined and repeating or periodic pattern, usually a wave function. In natural biological systems these occur as part of circadian clocks where the timing of gene expression follows a cyclical pattern. They also occur in the central nervous system which displays rhythmic patterns of neural activity and in many biochemical networks. An oscillator is a device that creates oscillations in output. Many schemes for synthetic oscillators have been explored on both a theoretical and experimental level because such devices are necessary in order to create complex systems. The goal in all cases is to create a system that results in stable, sustained and robust oscillations.

Oscillations are defined by two parameters: their periodicity and their amplitude (Fig. 5.6). The period of an oscillation is the length of time that elapses from peak output to peak output. This can also be thought of as the length of time that a single cycle takes. The frequency of an oscillation, or how many oscillations occur in a given time period, can be calculated by taking the inverse of the period. The amplitude of an oscillation is the distance between the average and the maximum output of the oscillation.

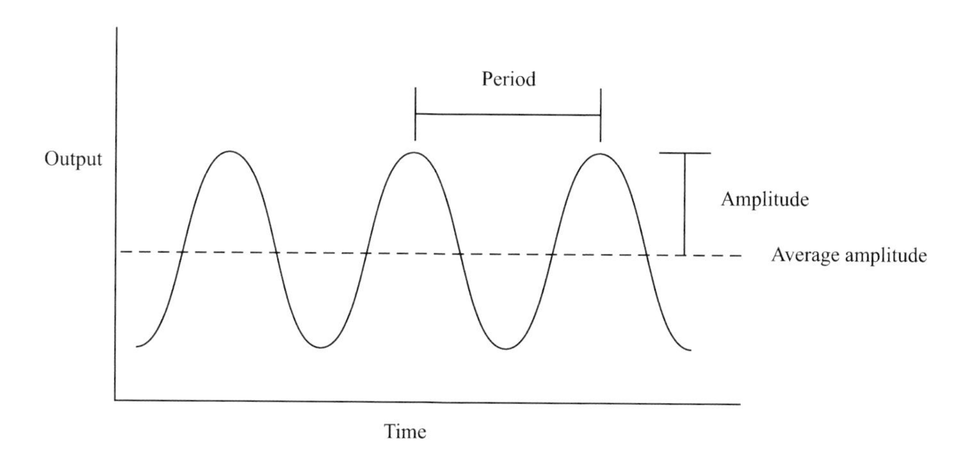

Fig. 5.6 A schematic of an oscillator.

5.4.3.1 *Single gene coupled with delay*

One generic design for an oscillator is to couple a negative feedback loop with a delay or lag phase. This design can be tuned to create system instability which will lead to oscillations. The simplest system design with this organisation is a single gene with negative feedback regulation of itself, coupled to a mechanism for delay. This is an organisation very similar to a delay line oscillator and has been demonstrated in theoretical studies.

5.4.3.2 *Lotka–Volterra oscillators*

Sender and receiver devices can be combined into two different cell populations to form an oscillator that is governed by the Lotka–Volterra equations. These are the same equations which govern the behaviour of predator–prey interactions in ecosystems. In a Lotka–Volterra oscillator, two different populations of cells are mixed, one which synthesises a signal at an exponential rate (prey), and one which causes this signal to decay exponentially (predator). In this instance, oscillations are induced by the time delay in response as signals are transmitted from one population to another, leading to a lag in response. This oscillator follows the same generic design of a negative feedback with delay. The predator consumption of the signal leads to negative feedback and the diffusion and processing of the signal molecules introduces the delay.

5.4.3.3 *Repressilators*

Repressilators are oscillators constructed by coupling together multiple inverters. The classical repressilator consisted of three inverters (Fig. 5.7), but theoretical investigations suggest that any number of odd inverters in series will produce sustained oscillations. Here again, the principle of a negative feedback loop with delay applies. The negative feedback comes through the use of an odd number of genes in a cyclic structure, while the delay is introduced by the temporal behaviour of the system along the cyclic gene structure. Repressilators have also been constructed that contain senders and receivers to coordinate the oscillations among a population of cells.

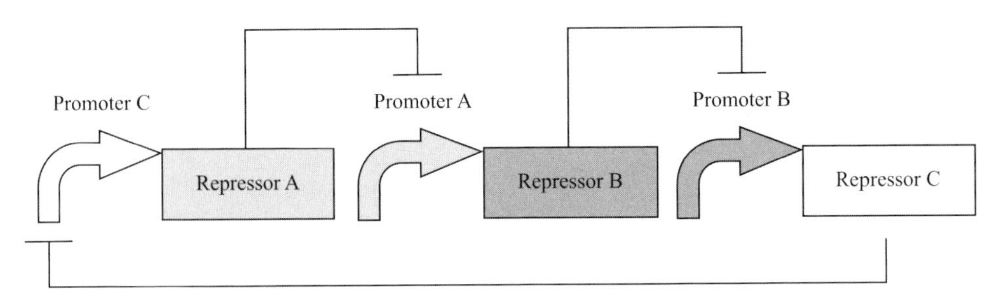

Fig. 5.7 A generalised repressilator composed of three inverters.

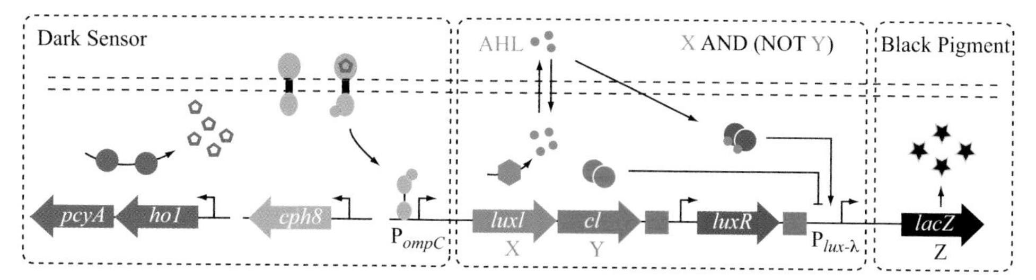

Fig. 5.8 A schematic of an edge detector. The circuit is composed of a dark-sensing element and an AHL receiver which together help cells to determine whether they are on the boundary of light and dark and, if so, make a black pigment. (Courtesy: Tabor *et al.*, 2009).

5.4.4 *Edge detector*

Edge detection algorithms are used in image processing to define the boundaries between objects by detecting shifts in the brightness when moving from pixel to pixel. In synthetic biology edge detectors are necessary when using light control elements to ensure that cells at the boundary between light and dark behave in a controlled manner. The edge detector described in Fig. 5.8 combines a light-control element with a sender, a receiver and two logic gates to create an edge detection algorithm. The edge detector system is an excellent illustration of the principles of synthetic biology whereby more complex behaviours are created from well-characterised components. This system was initially constructed for use with the bacterial camera, but is useful for any light-control scenario. In the edge detector, two logic gates are used to help the cells to determine whether or not they are at the boundary of light and dark. Cells which are in the dark (as determined by the light-sensing element within them as a NOT gate, i.e. NOT light) activate a sender device which produces AHL. Cells which are in the light do not produce anything, but act as receivers for the AHL molecule from neighbouring cells. Cells which are at the boundary are in the light, so don't produce AHL, and they also receive a signal from the cells next to them (which are in the dark) and are activated to produce a pigment. Hence, they follow the logic X (I receive the AHL) and NOT Y (I am in the light) and produce a pigment. This works because although the AHL can diffuse it is taken up by cells in close proximity to the dark before reaching the total population.

5.4.5 *Counters*

Systems have been produced which allow bacteria to record events (also called counters) which allows the population to retain a memory of past events. This can be accomplished by coupling protein generators that are dependent upon one another for function. For instance, if the first event causes the production of a non-native protein that is necessary for the activation of a second protein generator, then the second event can trigger that output only after the first has occurred. The sequence can be continued if the output of the second event can control the third, etc. However, one drawback to such schemes is that the circuits become very large because a new protein generator is necessary for every event that one wishes to count.

5.5 Connecting Parts and Devices

Connecting parts and devices into a unit is not necessarily a simplistic task as the biological context in which a part or device is contained can affect its output. What this means is that the nucleic acid sequences surrounding a part can affect how that part functions. For instance, a promoter function is primarily affected by its sequence, but is also secondarily affected by the DNA base pairs both upstream and downstream of it, which affect the melting temperature of the DNA and the secondary structure produced when the strands are parted for transcription to take place. Within organisms, this property can be used to fine-tune the output from various promoters. Similarly, when connecting different devices, the nucleic acid sequences in between will affect the function. This added layer of complexity must be taken into account when building complex systems.

5.5.1 *Tuning secondary structure of mRNA in and around parts*

When constructing biochemical networks to synthesise a product of interest, it is often necessary to tune the output of several protein-generating devices to optimise flux through the pathway and prevent intermediates from building up. One method to accomplish this is to tune the mRNA structure flanking different parts in the pathway so that the secondary structure of the mRNA balances translation. Examples include the DNA sequence of the promoter in between consensus binding sites, the region around the ribosome binding site and the intergenic regions between protein coding devices. While there are some algorithms that have been developed for this type of calculation (see below), another option to accomplish the same task is to create a combinatorial library which varies the secondary structure somewhat randomly and to screen for increased output from the device or system. Although this method is labour intensive, it provides the ability to experimentally explore the effect of variations in protein output on the overall yield from a system.

5.5.2 *RBS matching*

The output of protein generators has also been tuned by optimising the strength of the ribosome binding site computationally. An algorithm has been developed to predict the output based on the free energy of interaction between the 30S ribosomal complex and the ribosome binding site. The algorithm also takes into account the immediate context of the ribosome binding site, that is, the identity of the nucleic acids surrounding it. This helps to guide the design of metabolic pathways by providing information on the proportional output of a protein generator. This must still be paired with design information that suggests how much of each protein is necessary to optimise flux in the pathway.

5.5.3 *Insulators*

Work has also investigated the potential for insulating the interconnection of parts to remove the problems of context dependency. This could be accomplished by adding buffering sequences between parts in order to prevent the formation of adverse secondary structures that could impact the independent function of parts. However, questions remain as to how to experimentally implement this idea fully. For example, the ideal structure of a buffering region is itself context-dependent because it depends on the sequence around it. In addition, some parts must be closely spaced (e.g. a ribosome binding site and the start of a protein coding sequence), so that it will be impossible to insert buffers in all contexts.

A theoretical method of insulating the interconnection of two devices has also been proposed through the use of small circuits to prevent the devices from impeding each other. The ideal insulator would not interact with the upstream device, but simply receive the signal and generate a stable output independent of the identity of the downstream device. However, in order to perform an insulating function it is necessary to synthesise and degrade biomolecules at a very high and rapid rate in order to be able to generate a stable output. Thus, it is important to note that insulating devices will incur a very high metabolic burden on the cell.

5.6 Summary

The foundation of synthetic biology is a hierarchical assembly of smaller parts into devices into systems. This chapter has given a basic overview of some of the common parts, devices and small systems that have been constructed to date. For in-depth descriptions of the components discussed in this chapter, the reader is referred to the following research articles:

Reading

Toggle switch:

Gardner, TS, Cantor, CR and Collins, JJ. (2000). Construction of a genetic toggle switch in *Escherichia coli*. *Nature* 403: 339–342.

Oscillators:

Danino, T, Mondragón-Palomino, O, Tsimring, L, *et al.* (2010). A synchronized quorum of genetic clocks. *Nature* 463: 326–330.

Elowitz, MB and Leibler, S. (2000). A synthetic oscillatory network of transcriptional regulators. *Nature* 403: 335–338.

Stricker, J, Cookson, S, Bennett, MR, *et al.* (2008). A fast robust and tunable synthetic gene oscillator. *Nature* 456: 516–519.

Edge detector:

Tabor, JJ, Salis, HM, Simpson, ZB, *et al.* (2009). A synthetic genetic edge detection program. *Cell* 137: 1272–1281.

***RBS matching*:**

Salis, HM, Mirsky, EA and Voigt, CA (2009). Automated design of synthetic ribosome binding sites to control protein expression. *Nat Biotechnol* 27: 946–950.

***Insulators*:**

Del Vecchio, D, Ninfa, AJ and Sontag, ED (2008). Modular cell biology: retroactivity and insulation. *Mol Syst Biol* 4: 161.

CHAPTER 6

Modelling Synthetic Biology Systems

6.1 Introduction

In this chapter, we examine the role that mathematical modelling can play in the rigorous forward-engineering design of robust synthetic biology systems of increasing complexity. Synthetic biology aims to rationally construct complex biological systems from well-characterised components in the way that, for example, an electronic circuit may be designed. Engineers have long had to deal with some of the challenges that synthetic biologists are now confronted with. The use of mathematical models and of model-based computer aided design tools is widespread in other engineering disciplines and has enabled the efficient design and manufacture of complex systems composed of a large number of interacting parts.

6.2 Why is Mathematical Modelling Important and Useful?

As can be witnessed from other engineering disciplines such as chemical, mechanical or electrical engineering, the efficient design of complex systems can be greatly facilitated through the use of a rigorous forward-engineering approach rooted in mathematical modelling, model analysis and systems design and control. In this context, the first question one might want to ask is: What exactly is a model?

A model is a representation of the essential aspects of an existing system or of a system to be constructed which presents knowledge of that system in a usable form for its analysis or for its design. Models are thus not replicas of reality; they are simplified descriptions of it. It is important to understand that there is no such thing as 'the' model. A model can only be defined based on the type of questions that one seeks to answer, e.g. how can one design a gene regulation system that produces proteins whose intracellular concentrations oscillate periodically in time? Or how can we engineer a metabolic pathway allowing *E. coli* bacteria to produce as much ethanol as possible from a carbon source such as glucose? These questions determine the level of abstraction or granularity and type of model that needs to be built.

Building models requires both the specification of the model structure for the desired system (e.g. set of parametric equations describing the constituent parts and their interconnections) and the identification of the set of parameters appearing in the model (e.g. production and decay rates). It is also important to estimate biologically feasible ranges of variation of the model parameters within which the desired behaviour of the system is preserved.

Mathematical modelling is the result of a trade-off between accuracy and simplicity since although large, complex models may lead to more accurate results they typically require many more parameters whose values may not be easily accessible. Furthermore, the behaviour of very large models might also be harder to analyse/understand or to numerically simulate. Therefore, building 'good' models takes practice, experience and iteration. The goal of a 'good' model is to appropriately capture the fundamental aspects of the system of interest while leaving out the details that are irrelevant to the questions that are asked. With this goal in mind, the modelling process has to take into account the appropriate time and spatial scales that need to be considered, the type of data available, and also the types of simulation, analysis and design tools to be used. The modelling process is considered successful when the model obtained possesses the following characteristics:

Accuracy: the model should attempt to describe current observations with a 'sufficient' level of accuracy.
Predictability: the model should allow one to predict the behaviour of the system (through analysis or simulation) in situations not already observed.
Reusability: the model should be reusable in other similar cases.
Parsimony: the model should be as simple as possible. That is, given competing and equally good models, the simplest one should be preferred.

6.2.1 *What are the uses of mathematical modelling in synthetic biology?*

Mathematical modelling plays a crucial role in the efficient and rational design of robust and complex synthetic biology systems since it serves as a formal link between the conception and physical realisation of a biological circuit. In particular, the rational building of robust systems of increasing complexity from the interconnection of different parts or devices can be significantly facilitated through the use of a forward-engineering approach relying on the separation of the *in silico* model-based design from the actual *in vivo* or *in vitro* wet laboratory implementation. This approach allows various designs to first be tested and optimised *in silico* using model-based computer simulations and mathematical analysis methods before committing any effort or time to their *in vivo* or *in vitro* realisation.

When appropriately developed, 'good' mathematical models allow for design decisions to be taken regarding how to interconnect subsystems, choose parameter values and design regulatory elements so as to propose a system implementation satisfying the design specifications (e.g. desired behaviour, robustness to environmental perturbations, fundamental limits on system operation and performance).

6.3 Forward-engineering Approach to the Design of Synthetic Biology Systems

As the requirements of synthetic biological systems have become more complex, the need for modelling methods and software design tools has become more acute. To ensure that mathematical models can be efficiently reused in the design of increasingly complex systems, the developed models must be easily composable (e.g. by interconnecting them using well-defined inputs and outputs for each model) and the behaviour resulting from their interconnection must be predictable from the behaviour of the components and the way they are interconnected. This last step is the goal of systems and control theory, and its application to the analysis and design of biological systems has led to many recent developments. The definition of composable design models and of their appropriate interconnection is a key step in enabling the robust design of complex systems from the interconnection of several parts. This step is at the core of the forward-engineering approach of synthetic biology systems in which, *ideally*, the designer can use the following workflow (Fig. 6.1).

The design begins by the definition of the design objectives, i.e. what dynamical phenotypic behaviour is sought and with what properties (e.g. robustness, yield, time response); and under what constraints (e.g. upper limits on output variability, definition of chassis and environmental constraints). Based on these design specifications, different possible designs are envisioned. These designs differ by the choice of components or parts used, the values of the parameters appearing in the part models and the way these parts are interconnected.

For each design a set of mathematical models is constructed, ideally by using a standardised and curated database of composable models for the parts and considering the interconnection rules imposed by the considered design. Model-based analysis and simulations are performed to assess the performance of each design *in silico* with respect

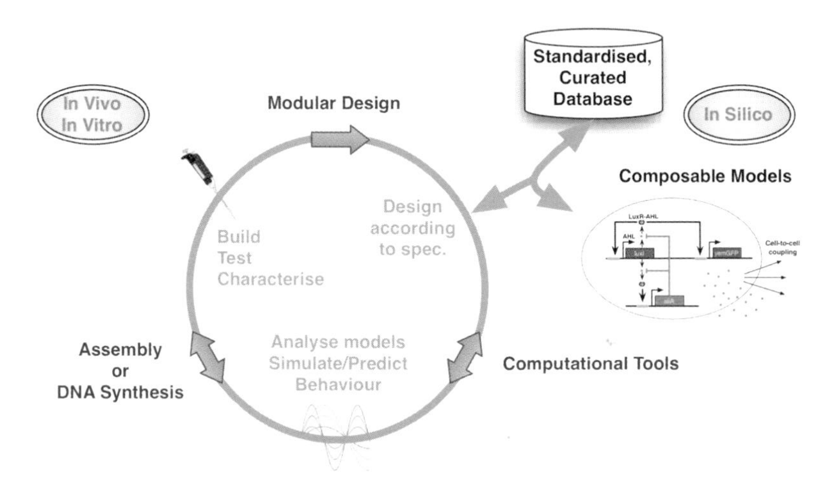

Fig. 6.1 The engineering design cycle — design starts *in silico* and proceeds iteratively along the cycle. The use of Computer Aided Design tools can allow for the *in silico* modelling, analysis and optimisation of the design prior to the actual wet laboratory implementation of the system. (Inspired by Chandran *et al.*, 2008.)

to the initial design specifications. At this stage, *in silico* analysis and optimisation allows one to search for parameter values leading to the desired behaviour. Furthermore, using robustness analysis, different models can be assessed with respect to their ability to withstand structural and dynamic perturbations. This typically leads to *in silico* iterations whose goal is to eventually select a subset of design candidates for the wet laboratory implementation.

The *in silico* candidate designs are then 'translated/compiled' into DNA sequences which are then implemented *in vitro* or *in vivo*. The implementation is tested and characterised to yield biological data that are then compared with the initial design objectives, thereby allowing assessment of the quality of the design and, if necessary, improvements by iterating once or several times around the design cycle. The use of Computer-Aided Design (CAD) tools can make iterations between these different steps easier and more efficient. Furthermore, CAD tools can be supplemented by Graphical User Interfaces (GUIs), which enable the construction of devices and systems by graphically interconnecting components on a canvas and automatically building the corresponding models in the background. Ultimately, computational tools could also be used to predict the DNA sequence that is required for the *in vivo* or *in vitro* implementation of the designed model into a particular host cell or even to control and automate the DNA assembly or *de novo* synthesis process using liquid-handling robots.

The engineering design cycle shown in Fig. 6.1 enables the efficient design of synthetic biology systems of increasing complexity using a forward-engineering approach very similar to the one successfully used in many other engineering disciplines. In particular, this approach enables the adoption of another important engineering principle: abstraction (see Chapter 3). Abstraction allows the construction of complex systems without the need for a detailed understanding of the processes involved at each stage along the design cycle. In other words, using abstraction each step along the cycle can be realised by different experts who do not need to know the details of the implementation of the other steps. The only knowledge required is that of the interface between the steps, what each step receives from the others and what it provides to the others. In particular, using abstraction system design and system fabrication can be separated (Fig. 6.2).

The advantage of such a separation is that it saves time, money and effort since the main burden of the design is realised more efficiently *in silico* than *in vivo* or *in vitro*.

6.4 Types of Mathematical Models and their Role

In the following sections we briefly introduce various types of mathematical models that are typically used in synthetic and systems biology to capture the essential dynamics of biochemical processes.

6.4.1 *Choosing the level of mathematical abstraction/resolution of the model*

As in other disciplines, synthetic biology systems can be modelled in a variety of ways and at many different levels of resolution and time scales (Fig. 6.3).

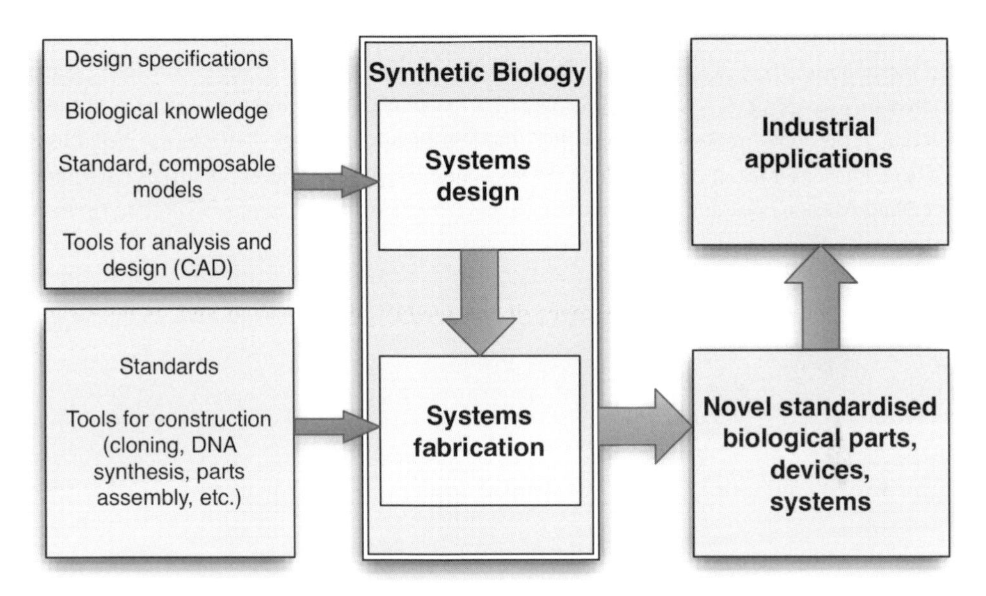

Fig. 6.2 Using the engineering design cycle in Fig. 6.1 and the concept of abstraction, system design can be separated from system fabrication. (Inspired by Heinemann and Panke, 2006.)

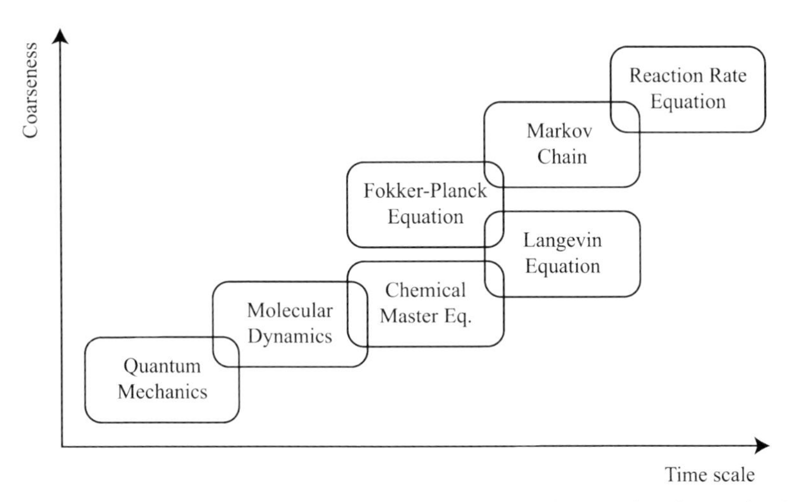

Fig. 6.3 Levels of abstraction typically used in the modelling process (Courtesy: http://www.cds.caltech.edu/~murray/amwiki/index.php/Supplement:_Biomolecular_Feedback_Systems.)

For example, we can try to model the molecular dynamics of the components of the cell, in which case we attempt to model the individual proteins and other species and their interactions via molecular-scale forces and motions. At this scale, the individual interactions between proteins, nucleic acids and other biomolecules are resolved, resulting in a highly detailed/complex model of the dynamics of these subunits.

Most of the time, however, such a level of resolution is both computationally intractable and too quantitatively inaccurate (due to the lack of knowledge of many parameter values) to answer the questions that one is interested in during the *design* of synthetic biology systems. Therefore, more coarse-grained models using Ordinary Differential Equations (ODEs), Partial Differential Equations (PDEs), Stochastic Differential Equations (SDEs) or Chemical Master Equations (CMEs) are typically used to model simple synthetic biology circuits (Fig. 6.4).

6.4.1.1 *Temporal (ODEs) versus spatio–temporal (PDEs) models*

Two coarse-grained types of model typically used are Ordinary Differential Equations (ODEs) and Partial Differential Equations (PDEs). These coarse-grained models can be

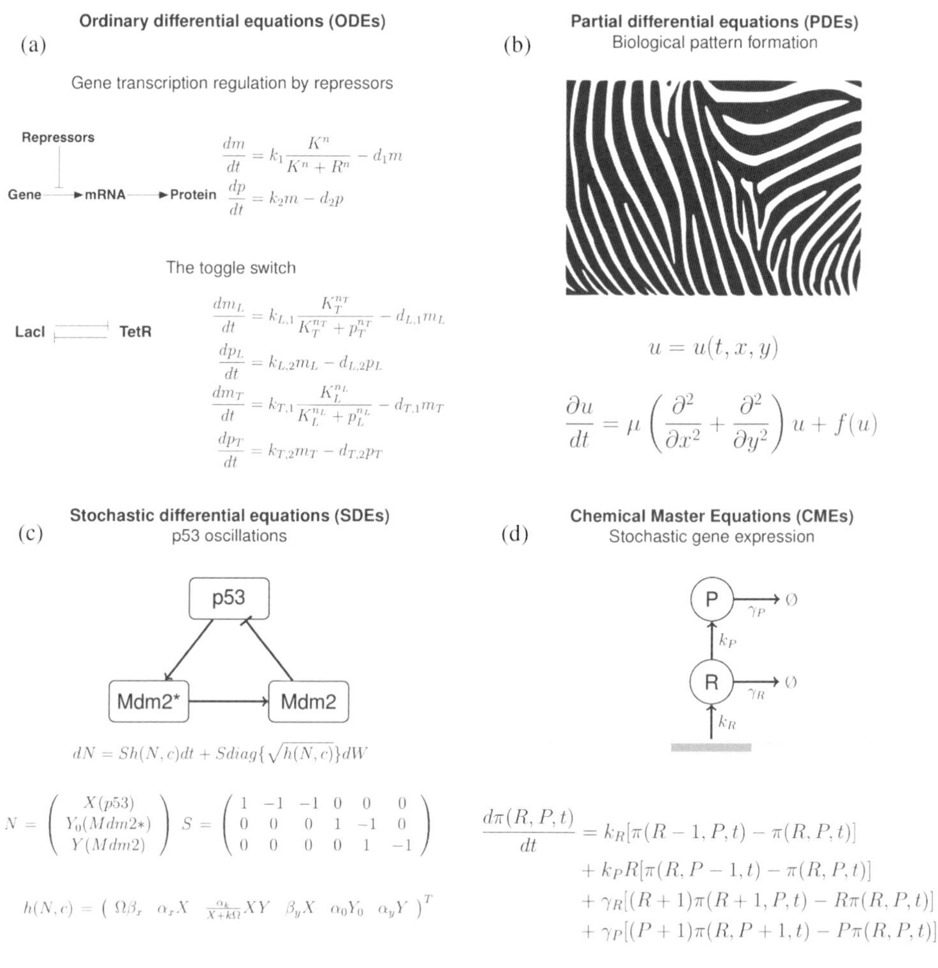

Fig. 6.4 *(Continued)*

used as simplifications of the real processes as long as their corresponding assumptions are satisfied. One typical assumption, for example, is spatial homogeneity (either within the cell or at the population level). Under the assumption of a spatial homogeneity, ODEs models are most commonly used. In ODEs, each variable, for example biochemical species concentration, can only depend on time but not on space (see Fig. 6.4a for example, where the variables $p(t)$ and $m(t)$ are only function of time). On the contrary, if spatial variations or non-homogeneities need to be explicitly taken into account, then the modelling may require the use of PDEs where the variables can depend both on time and on space (see Fig. 6.4b where the variable $u(t, x, y)$ is a function of time and of the spatial coordinates x and y. In this model we only consider two spatial coordinates, i.e. u is assumed to be evolving on a plane).

ODEs and PDEs are widely used in synthetic biology modelling. For example, some of the first synthetic biology systems such as the toggle switch and the repressilator were modelled using ODEs. On the other hand, to capture spatial variations in the dynamics of a population of bacteria acting as synchronised oscillators, PDEs were used.

Fig. 6.4 The different types of models typically used in systems and synthetic biology. (a) Using ordinary differential equations to model gene transcription regulation by repressors. The model possesses two variables $m(t)$ and $p(t)$. $m(t)$ represents the concentration of mRNA obtained through transcription of the considered gene while $p(t)$ represents the concentration of protein obtained through translation of the mRNA, at time t. The proposed model contains several parameters: the maximal transcription rate k_1, the repression coefficient K, the Hill coefficient n, the mRNA degradation rate d_1, the protein translation rate k_2 and the protein degradation rate d_2. In this model, R is considered as an external input representing the concentration of transcriptional repressor while the model output is chosen to be the amount of protein produced. Based on this simple model for gene transcription regulation by repressors a more complex model for two mutually repressing genes can be proposed. In this model the protein produced as the result of the expression of one gene acts as a repressor of the transcription of the other gene and reciprocally. This latter construct is called the toggle switch and was modelled and built by Gardner, Cantor *et al.* 2000. (b) Using partial differential equations to model biological pattern formation using the reaction–diffusion equation. The model considered here possesses a single variable $u(t, x, y)$ which depends on time t but also on the planar spatial coordinates x and y. The change in local concentration of each chemical species over time is a function of the Laplacian of the local concentration (to account for diffusion) and the local concentration of the chemical species (to account for chemical reactions) (Turk, 1991; Meinhardt and de Boer, 2001). The parameter μ is the diffusion coefficient (a measure of how fast molecules diffuse from regions of high concentration to regions of low concentration) while the function $f(u)$ is defined based on the specific details of how the molecules in the system react with each other. (c) Stochastic differential equations (based on the Langevin approximation) for the p53 oscillation model (Geva-Zatorsky, Rosenfeld *et al.*, 2006), where dW is the increment of the Wiener process (also known as Brownian motion — the stochastic element of the equation) and X, Y_0 and Y represent the numbers of p53, Mdm2 precursor and Mdm2 respectively. Ω represents the volume of the system and α, β and k represent the production and degradation rates. As with the previous types of models described above (a, b and c), the variables X, Y_0 and Y vary continuously (i.e. they cannot take discrete values). (d) Chemical master equation for a simple stochastic gene expression model (Thattai and van Oudenaarden, 2001) where R and P represent the number of RNA and protein molecules respectively, and $\pi(R, P, t)$ is the probability of observing R RNA and P protein molecules at time t. In contrast to the previous types of equations, R and P are discrete integer values and change stochastically in discrete jumps over time according to the parameters of the model. The parameters k_R and k_P are the production rates of mRNA and protein, while γ_R and γ_P are degradation rates of mRNA and protein respectively.

6.4.1.2 *Stochastic (SDEs) versus deterministic (ODEs) models*

Another important distinction occurs based on the type of modelling framework used, deterministic versus stochastic. The deterministic framework is appropriate to describe the mean behaviour (averaged across a large number of molecules) of a biochemical system. Deterministic models implicitly assume that the underlying variables such as concentrations or molecule numbers vary in a deterministic and continuous fashion. In other words, there is no element of randomness associated with the corresponding model, and thus given the same initial conditions (e.g. same initial concentrations) the time evolution of all the variables is always the same. On the other hand, the stochastic framework takes into account the random interactions of biochemical species. More specifically, stochastic models are used to mathematically capture stochastic variations and noise, two properties inherent to biological systems. They are typically used when the number of species involved is 'small' and stochastic effects can no longer be 'averaged out'. This is, for example, the case for transcription factors, which, in certain circumstances, can be expressed at low levels (a few tens of molecules), or for DNA, for which a single copy may exist in the cell. Due to the presence of some randomness in stochastic models, the time evolution of variables will not be necessarily the same for a given model simulated several times starting from the same initial conditions.

Two main types of stochastic models — Stochastic Differential Equations (Fig. 6.4c) and Chemical Master Equations (Fig. 6.4d) — are typically used to represent stochastic systems. The analysis of stochastic models can occasionally be realised by mathematically deriving the most important statistical moments (such as mean and variance), although this can be quite hard for complex models such as those typically found in realistic biological modelling. Therefore, understanding the behaviour of complex stochastic models is typically done through large numbers of computer simulations.

Simulation of CMEs models is very computationally expensive, especially if the number of different biochemical species in the model is large, the reaction rates are high or the number of molecules of each species increases in size. Simulation from SDEs is less computationally expensive but SDEs do not accurately represent chemical processes at (very) low number of molecules. For this reason, there has been an effort to develop hybrid models that combine deterministic (ODE) and stochastic (SDEs and/or CMEs) modelling into a single framework where each species is modelled most appropriately while minimising computational time.

Since designs that specifically use stochastic aspects as features are quite hard to obtain, one typically starts with deterministic designs/models and, if necessary, the behaviour and performance of these designs are then put to the test by creating corresponding stochastic models and checking that the stochastic aspects do not significantly impact on the initially designed behaviour or performance.

6.4.2 *Experimental data, model resolution and parameter identification*

Choosing the appropriate level of resolution of a model (see Fig. 6.3) also requires taking into account the type, amount and quality of experimental data that are or can be made available. Indeed, a model with a very high resolution might not be very useful if the amount and quality of data available to estimate the model parameters is low or if these data do not really allow one to assert the level of confidence one can have on the created model (to accept or invalidate the designed model by comparing its predictions with measured data). This also raises the question of identification of model parameters from the data, i.e. what is the minimal amount of information necessary in order to be able to estimate the parameters of a given model. This question is not only important for the appropriate design of models but also for the design of experimental protocols targeting the extraction of the minimal information required to address the parameter identification problem.

6.5 Model Repositories

In order to facilitate the modelling process, several biological model repositories have been created in recent years. These repositories are accessible online and serve as a centralised source of information that allows users to store, search and retrieve published mathematical models of biological interests. A non-exhaustive list of biological model repositories can be found at http://systems-biology.org/resources/model-repositories/. Among these model repositories, one of the most used is BioModels (http://www.ebi.ac.uk/biomodels-main/). It provides access to published, peer-reviewed, quantitative models of biochemical and cellular systems. The models on BioModels are stored under the form of biochemical reactions which can be used to automatically create both deterministic (e.g. ODEs) or stochastic (e.g. SDEs) mathematical models.

It is important to note that each model in the BioModel repository is carefully curated to verify that it corresponds to the referenced publication and gives the proper numerical results. Curators also annotate the components of the models with terms from controlled vocabularies and links to other relevant data resources. The models published on the BioModels website can be simulated online (deterministically or stochastically) and exported into various formats including the popular Systems Biology Markup Language (SBML) (http://sbml.org/Main_Page) or the CellML format (http://www.cellml.org/). These are machine-readable formats for representing computational models at the biochemical reaction level. By supporting SBML or CellML as input and output formats, different software tools can operate on the same representation of a model, removing chance for errors in translation and assuring a common starting point for analyses and simulations.

6.6 Information Necessary for Synthetic Biology Modelling

As we saw in the previous section, biological models typically contain several parameters whose values influence the time evolution/behaviour of the considered model (for example, k_1, k_2, K, n, d_1 and d_2 in the ODEs model for gene transcription regulation by repressors of Fig. 6.4a). So even if, based on first principles or engineering design considerations, the structure of the system can be specified under the form of a parametric model, a careful estimation of the model parameters will be required to ensure the appropriate behaviour of the designed system. In the next three sections, we give an overview of the sources of quantitative information that are useful to identify model parameters.

6.6.1 *Datasets from the biological literature*

Parameter values are typically inferred from measured experimental data (see 'Parameter Inference and Model Selection'). Sometimes parameter values can also be derived from biophysical values using insight into the biological processes involved. Most often, the first step when looking for particular parameter values consists of searching the specialised literature and sifting through a large amount of papers to extract the relevant information. This, however, can be quite time-consuming and inefficient. This is why scientists and researchers have started gathering and curating this information in order to deposit it in freely accessible, online repositories. One such popular repository recently developed at the Systems Biology Department in Harvard (USA) and the Weizmann Institute (Israel) is called BioNumbers (http://bionumbers.hms.harvard.edu/). The BioNumbers website provides access to a growing number of biological and biophysical values extracted from the literature, together with the source used to provide the values. The centralisation of this information greatly facilitates the search for biologically relevant values, which can be useful for estimating the values of parameters appearing in biological models.

6.6.2 *Datasheets from the characterisation of synthetic biology parts*

Another way of obtaining useful information for the design of models and the identification of their parameters comes from the direct and methodical wet laboratory characterisation of engineered synthetic biology parts. Electronic engineers use TTL Data Books listing standard electronic components and their characteristics in order to choose the components required to construct their circuits. In a similar way, the ultimate goal of standardisation and characterisation of synthetic biology parts is to provide synthetic biologists with a catalogue of standard parts together with their associated characterisation datasheets. To be useful, these datasheets should contain enough quantitative information to allow a synthetic biologist to appropriately choose the specific parts that will yield the desired behaviour for an engineered system. An example of 'gold standard' datasheet is the BBa_F2620 part datasheet giving quantitative characterisation information of a genetic receiver device. This receiver device takes the concentration of an inducer molecule as input and produces a specific output signal in response. The corresponding datasheet defines a number of aspects of the behaviour of the device: the strength of the output signal at different inducer concentrations, the time evolution of the output, the reliability or expected time to failure of the device, the degree of compatibility with other inducers and an indication of the type and amount of cellular resources the device consumes. Significantly, the authors of this datasheet underline the importance of measuring a large number of properties in order to comprehensively characterise genetic elements, especially those that describe the way the parts interact with the chassis in which they are intended to function. This hints at the necessity of producing high-quality datasheets of well-characterised parts in order to support the rapid and efficient reuse of the engineered parts or devices in various systems of increasing complexity. The creation of a registry of professionally characterised biological parts is one of the main goals of BIOFAB (see Chapter 3). These characterisation data are very useful to build mathematical models of the characterised parts and to identify their parameters.

6.6.3 *Parameter inference and model selection*

Given experimental data of sufficient quality and quantity it is possible to estimate the values of parameters appearing in mathematical models (this is called parameter inference or parameter estimation) and to decide which model is 'better' at explaining a given process (this is called model selection).

Typically, wet laboratory experimental data consist of time course measurements of the model species under some *in vivo* or *in vitro* conditions, such as time series measurements of protein concentrations under specific experimental conditions. Given these experimental data, parameter inference can be posed as a parameter optimisation problem where the function to be optimised takes as inputs the experimental data and the model with the chosen parameter values and gives as output a measure of how well simulations/predictions of the model with these particular parameter values match the experimental data. The optimal parameter estimates are those that maximise the objective function.

Model selection, on the other hand, refers to the task of selecting the 'most appropriate' models within a set of competing candidate models. Generally, models with more parameters fit the data better than models with fewer parameters and this must be accounted for. This is typically done using a parsimony principle so that given two competing models which explain the data equally well, the simpler one will be preferred. This is done using concepts such as the Likelihood Ratio Test (LRT), the Akaike Information Criterion (AIC) or the Bayesian Information Criterion (BIC).

6.7 Information Provided by Synthetic Biology Models

6.7.1 *Metrics of interest in synthetic biology*

From an engineering design point of view, the purpose of mathematical models in synthetic biology is to enable fast, efficient and reliable prediction and pre-assessment of the behaviour and performance of a designed system before its actual wet laboratory implementation. In particular, based on a mathematical model several metrics of interest to the designer (see Fig. 6.5) can be estimated *in silico* for the envisioned design(s). These metrics can be, for example, the time it takes for the system to respond to a step change at its input, the expected steady-state amplitude of the output of the system, the tunability of the output with respect to some of the parameters of the model or the robustness of the system to environmental perturbations. All of these metrics are extremely important in the appropriate design of a synthetic biology system. Estimating them *in silico* based on constructed models can be done efficiently through mathematical analysis and numerical simulations of the models.

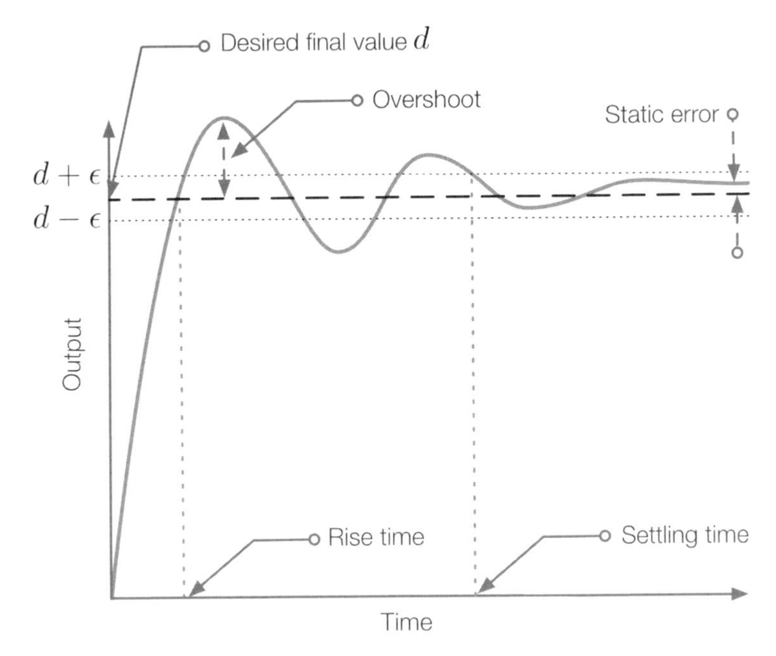

Fig. 6.5 Example of typical performance metrics for the time evolution of the output of a designed system.

6.7.2 *Model-based numerical simulations*

To ensure that the *in silico* designed systems behave as desired, the proposed designs and their corresponding models are numerically simulated in various situations. This step is useful to check that models are characterised by behaviours and performances coherent with the initial design specifications. During this step the various metrics of interests are estimated *in silico* (Fig. 6.5). If the behaviour or performance are not satisfactory then some additional *in silico* iterations will be necessary to refine the designs/models until they are deemed acceptable.

6.7.3 *Fundamental limits on system operation*

Mathematical analysis of a model provides the designer with important information regarding the possible types of behaviours that can be expected from the implementation of a model, and also the fundamental limits of the designs, as well as ways to improve the design.

6.7.3.1 *Identifying important controlling factors and weak spots in the design analysis of the models*

Once models have been created and their parameters determined, these models can be mathematically analysed in order to characterise their qualitative behaviour and identify important controlling factors and ways to improve the design. Which parameters have the 'largest impact' on the model behaviour? This is typically discovered through what is called a parameter sensitivity analysis explained in more detail below. What types of behaviours can be expected from the considered model and how can this behaviour change when the parameter values are modified? This is done through bifurcation analysis (see below). How robust is the behaviour/performance of the designed model to structural and dynamic perturbations?

6.7.3.1.1 Parameter sensitivity analysis

The goal of a parameter sensitivity analysis is to determine the influence of each parameter or combination of parameters on the behaviour or performance of the model. Sensitive parameters are those which are primarily responsible for significant changes in dynamical behaviour or performance and their value should be chosen wisely. Local sensitivity analysis is the study of the changes in the model output or performance function with respect to small perturbations in individual specified parameter values. Global sensitivity analysis enables an examination of the response of the system output or performance to simultaneous variation in parameter values across the whole admissible range of parameter variations. Global sensitivity analysis is of particular importance in biological models for which parameters can vary within large intervals depending on their meaning.

Examples of sensitivity analysis software tools developed specifically for biological systems include BioSens (http://www.thedoylegroup.org/research/BioSens/BioSens.htm), SensSB (Rodriguez-Fernandez and Banga, 2010) and SBML-SAT (Zi, Zheng *et al.*, 2008).

6.7.3.1.2 Bifurcation analysis

This type of analysis informs the designer about transitions in dynamical regime induced by variations in parameter values. In particular, bifurcation analysis allows one to characterise regions in the parameter space in which the steady-state behaviour of the system remains qualitatively the same. At the boundaries of such regions, a qualitative change in the steady-state behaviour occurs, e.g. increasing the value of a single parameter while keeping the others constant leads to a transition from a unique asymptotically stable equilibrium point to an asymptotically stable periodic oscillation. The characterisation of these regions provides crucial information for the design as it allows the identification of the parameter regions/ranges within which a desired behaviour can be obtained. Numerical continuation analysis tools such as XPPAUT (Ermentrout *et al.*, 2003) or MATCONT (Dhooge *et al.*, 2003) can be used to perform a bifurcation analysis of a given model once an initial guess for its associated steady-state solution is known.

6.7.3.1.3 Robustness analysis

Although several designs can lead to the same behaviour, their respective ability to withstand perturbations may be very different. Provided that enough information is known about the perturbation (at least an upper bound on the energy of the perturbation), robust analysis of the models allows the assessment of their relative immunity to such perturbations. In particular, this step can prove very informative for the robust design of systems, and established multivariable robust control techniques can be used to that effect in the design process.

A widely used tool for this type of analysis is the Matlab Robust Control Toolbox (http://www.mathworks.com/products/robust/). The SBML-SAT tool mentioned above also implements algorithms for robustness analysis, as does the BIOCHAM (Rizk, Batt *et al.*, 2009) modelling environment.

6.7.3.2 *Determining constraints/limits on system behaviour/performance through mathematical analysis*

By mathematically analysing a model or a family of models, some information can be gained regarding the conditions required to obtain a particular behaviour or ways to improve the design but also, and most importantly, regarding the fundamental limits that constrain the considered design — for example, by using a combination of control and information theory to mathematically obtain a fundamental limit on the maximum suppression of molecular fluctuations that can be obtained using *any* type of feedback control mechanism. The result shows that the limit is inversely proportional to the quartic root of the number of signalling events, making it extremely expensive to increase accuracy (e.g. to reduce molecular fluctuations by one order of magnitude you need to increase the number of signalling events by four orders of magnitude). This type of result is very useful for the design of synthetic biology systems since it informs the designer about what can be expected at best from any design and thus allows one to define the limits of what is achievable by a particular design.

6.8 Computational Tools for *In Vivo* or *In Vitro* Implementation of Synthetic Biology Systems

Once mathematical analysis and numerical simulations indicate that the models satisfy the design specifications, the synthetic biologist can focus his or her attention on the laboratory realisation of the proposed designs. This initiates the construction, characterisation and testing phase of the engineering design cycle (Fig. 6.1).

The laboratory implementation requires translation of the candidate synthetic biology design(s) into a biological language that the host cell can understand and 'execute'. This typically means converting the design into an appropriate DNA sequence that the cell transcription/translation machinery can use to implement the desired synthetic biology circuit.

If the design is based on the interconnection of standard, well-characterised parts for which the associated models and parameter values are known then the implementation can be quite straightforward. In this situation, the physical design requires the assembly of the DNA sequences associated with the different parts into a larger section of DNA that implements the design (see Chapter 3). The assembly phase also requires the integration of this larger piece of DNA into plasmid vectors or into the chromosome of the host cell. The assembly phase can range in scale from assembling a gene optimised to efficiently express a protein with a given amino acid sequence, to assembling a set of genes to form a pathway, and up to assembling an entire bacterial genome. In this context, computational algorithms can help eliminate human error as well as minimise the time and cost of experiments with the ultimate aim of high-throughput fully automated assembly by liquid-handling robots.

6.8.1 *Computational tools for the automatic implementation of designs based on existing biological parts*

The laboratory implementation of a synthetic biology system becomes more complex if one does not really know which parts are needed to build the desired system. In such a situation, tools that (based on a specified model/behaviour) enable either automatic selection of parts or *de novo* synthesis of DNA sequences are necessary.

Some currently developed tools such as the GEC programming language (Phillips and Cardelli, 2009) can take a high-level description of the desired specifications/behaviour together with a database of standard biological parts with associated properties, and output a set of candidate designs (based on the appropriate choice of parts from the part database) that satisfy the requirements defined in the GEC program. The resulting set of candidate part-based designs can then be numerically simulated for further analysis and selection before the actual experimental implementation.

Another example is ClothoCAD (http://www.clothocad.org/). ClothoCAD is an ambitious large-scale project to develop a modular and integrated software platform for synthetic biology systems providing graphical sequence editing, data management, algorithm management tools and a plug-in system. This software includes the ability to interface with liquid-handling robots for automated production of the designed constructs.

6.8.2 *Computational tools for the implementation of designs requiring de novo DNA synthesis*

Automatic DNA sequence compilers are currently being developed to help synthetic biologists automatically predict the specific DNA sequences that need to be inserted into a chassis cell in order to implement the synthetic biology system that was designed *in silico*. In particular, current research efforts are focusing on developing calculators or DNA sequence predictors for the most fundamental synthetic biology parts, including promoters, ribosome binding sites, protein coding sequences and terminators. This kind of tool is extremely useful as it provides a link between some high-level property and its low-level implementation within the biomolecular language of the cell. In the future, such tools will provide the currently missing link, allowing the *de novo* prediction/compilation of DNA sequences required to implement any particular synthetic circuit.

6.9 Comparison of Model Predictions/Simulations with *In Vivo/In Vitro* Data

In order to close the engineering design cycle and to assess the quality of a model and its utility in the design of a synthetic biology system, the predictions of the model need to be

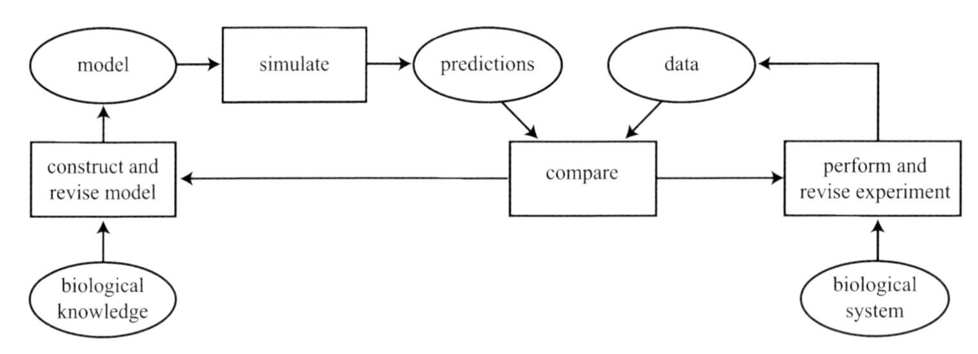

Fig. 6.6 Comparing model predictions with experimental data. Models are typically the results of several iterations aimed at refining the model based on these comparisons.

compared with experimental data obtained from the *in vivo* or *in vitro* implementation of the system (Fig. 6.6).

In the design context, *in vivo* or *in vitro* testing and characterisation of the synthetic biology construct allows one to feed information back about the behaviour and performance of the implemented systems, allowing for model refinement where necessary. Comparison with the initial design specifications then allows the designer to decide if the implemented systems satisfy the design requirements or if some additional iteration along the design cycle is necessary. Furthermore, predictions of the model may also be informative for the design of new experiments whose goal is to reveal previously unobserved behaviours of the system, thereby deepening its understanding.

6.10 Model and/or Experimental Refinements

If the synthetic biology systems implemented *in vivo* or *in vitro* do not satisfy the required design specifications or do not behave as predicted by the model, it is necessary to identify the cause of such discrepancy. Several causes of discrepancies must be considered: (i) the model is inappropriate and needs to be revised; (ii) the design specifications are too stringent and need to be relaxed; (iii) the implementation of the synthetic biology construct contains errors that need to be identified and corrected. The first two causes (i) and (ii) launch the synthetic biologist around another full iteration of the engineering design cycle (Fig. 6.1). (iii) requires the use of circuit debugging tools, such as DNA sequencing, and methods for biochemical network reconstruction from data to test if the (dynamical) structure of the implemented synthetic biology circuit corresponds to the one originally intended by its designer.

The approach and concepts that we have briefly introduced in this chapter constitute the main toolbox that engineers typically use to successfully build complex systems by relying on a rational approach based on mathematical modelling, model analysis and *in silico* simulations prior to actual system implementation. Taking a similar approach in synthetic biology will provide the community with foundational mathematical and computational tools required for the efficient forward-engineering of robust synthetic biology systems of increasing complexity.

Reading

Forward engineering:

Endy, D. (2005). Foundations for engineering biology. *Nature* 438: 449–453.

Systems and control theory:

Alon, U. (2007). *An Introduction to Systems Biology: Design Principles of Biological Circuits.* Chapman & Hall/CRC Mathematical and Computational Biology Series, Boca Raton, FL.
Iglesias, PA and Ingalls, BP. (2010). *Control Theory and Systems Biology*. MIT Press, Cambridge, MA.
Kitano, H. (2001). *Foundations of Systems Biology*. MIT Press, Cambridge, MA.
Wellstead, P, Bullinger, E, Kalamatianos, D, *et al.* (2008). The rôle of control and system theory in systems biology. *Annu Rev Control* 32(1): 33–47.

Engineering design cycle:

Chandran, D, Copeland, WB, Sleight, SC, *et al.* (2008). Mathematical modeling and synthetic biology. *Drug Discov Today: Disease Models* 5(4): 299–309.

Use of ODEs and SDEs in synthetic biology modelling:

Danino, T, Mondragón-Palomino, O, Tsimring, L, *et al.* (2010). A synchronized quorum of genetic clocks. *Nature* 463: 326–330.
Elowitz, MB and Leibler, S. (2000). A synthetic oscillatory network of transcriptional regulators. *Nature* 403: 335–338.
Gardner, TS, Cantor, CR and Collins, JJ. (2000). Construction of a genetic toggle switch in *Escherichia coli*. *Nature* 403: 339–342.
Kaznessis, YN. (2009). Multiscale models for synthetic biology. *Conf Proc IEEE Eng Med Biol Soc* 6408–6411.
Purcell, O, Savery, NJ, Grierson, CS, *et al.* (2010). A comparative analysis of synthetic genetic oscillators. *J R Soc Interface* 7(52):1503–1024.
Van Kampen, NG. (2007). *Stochastic Processes in Physics and Chemistry*. Elsevier, Amsterdam.
Wilkinson, DJ. (2009). Stochastic modelling for quantitative description of heterogeneous biological systems. *Nat Rev Genet* 10(2): 122–133.

Model resolution and parameter identification:

Gonçalves, J and Warnick, S. (2008). Necessary and sufficient conditions for dynamical structure reconstruction of LTI networks. *IEEE Trans Automat Control* 53(7): 1670–1674.
Ljung, L. (1999). *System Identification: Theory for the User*, 2nd Ed. Prentice Hall, Upper Saddle River, NJ.
Sontag, ED. (2008). Network reconstruction based on steady-state data. *Essays Biochem: Systems Biology* 45: 161–176.
Sontag, ED, Kiyatkin, A and Kholodenko, BN. (2004). Inferring dynamic architecture of cellular networks using time series of gene expression, protein and metabolite data. *Bioinformatics* 20(12): 1877–1886.

Yuan, Y, Stan, G-B, Warnick, S, *et al.* (2010). Network structure reconstruction from noisy data. *49th IEEE CDC*, Atlanta, Georgia.

Bionumbers:

Milo, R, Jorgensen, P, Moran, U, *et al.* (2010). BioNumbers — the database of key numbers in molecular and cell biology. *Nucleic Acids Res* 38(Database issue): D750–753.

Datasheets for biological parts:

Canton, B, Labno, A and Endy, D. (2008). Refinement and standardization of synthetic biological parts and devices. *Nat Biotechnol* 26(7): 787–793.

Model selection:

Burnham, KP and Anderson, DR. (2008). Model selection and multimodel inference. *Technometrics* 45: 18.

Parameter sensitivity analysis:

Kiparissides, A, Kucherenko, SS, Mantalaris, A, *et al.* (2009). Global sensitivity analysis challenges in biological systems modeling. *Ind Eng Chem Res* 48(15): 7168–7180.
Kitano, H. (2004). Biological robustness. *Nat Rev Gen* 5(11): 826–837.
Lestas, I, Vinnicombe, G and Paulsson, J. (2010). Fundamental limits on the suppression of molecular fluctuations. *Nature* 467: 174–178.
Saltelli, A. (2008). *Global Sensitivity Analysis: The Primer*. John Wiley, Chichester.
Saltelli, A, Ratto, M, Tarantola, S, *et al.* (2005). Sensitivity analysis for chemical models. *Chem Rev* 105(7): 2811–2827.
Zhou, K, Doyle, JC and Glover, K. (1995). *Robust and Optimal Control*. Prentice Hall, Upper Saddle River, NJ.

General References:

Del Vecchio, D and Murray, R. (2010). Biomolecular Feedback Systems. Available at: http://www.cds.caltech.edu/~murray/amwiki/index.php? [Accessed 6 September 2011].
Dhooge, A, Govaerts, W, Yu, A, *et al.* (2003). MATCONT: a MATLAB package for numerical bifurcation analysis of ODEs. *ACM Trans Math Software* 29(2): 141–164.
Ermentrout, B. (2003). Simulating, analyzing, and animating dynamical systems: a guide to XPPAUT for researchers and students. *Appl Mech Rev* 56(4): B53–B53.
Geva-Zatorsky, N, Rosenfeld, N, Itzkovitz, S, *et al.* (2006). Oscillations and variability in the p53 system. *Mol Syst Biol* 2: 2006 0033. Available at: http://www.nature.com/msb/journal/v2/n1/pdf/msb4100068.pdf [Accessed 5 November 2011].
Heinemann, M and Panke, S. (2006). Synthetic biology — putting engineering into biology. *Bioinformatics* 22(22): 2790–2799.
Lestas, I, Vinnicombe, G and Paulsson, J. (2010). Fundamental limits on the suppression of molecular fluctuations. *Nature* 467: 174–178.
Meinhardt, H and de Boer, PA. (2001). Pattern formation in *Escherichia coli*: a model for the pole-to-pole oscillations of Min proteins and the localization of the division site. *Proc Natl Acad Sci U S A* 98(25): 14202–14207.

Phillips, A and Cardelli, L. (2009). A programming language for composable DNA circuits. *J R Soc Interface* 6 Suppl 4: S419–436.

Purnick, PEM and Weiss, R. (2009). The second wave of synthetic biology: from modules to systems. *Nat Rev Mol Cell Bio* 10(6): 410–422.

Rizk, A, Batt, G, Fages, F, *et al.* (2009). A general computational method for robustness analysis with applications to synthetic gene networks. *Bioinformatics* 25(12): 169–178.

Rodriguez-Fernandez, M and Banga, JR. (2010). SensSB: a software toolbox for the development and sensitivity analysis of systems biology models. *Bioinformatics* 26(13): 1675–1676.

Salis, HM, Mirsky, EA and Voigt, CA. (2009). Automated design of synthetic ribosome binding sites to control protein expression. *Nat Biotechnol* 27(10): 946–950.

Stricker, J, Cookson, S, Bennett, MR, *et al.* (2008). A fast, robust and tunable synthetic gene oscillator. *Nature* 456: 516–519.

Strogatz, S. (1994). *Nonlinear Dynamics and Chaos: with applications to physics, biology, chemistry, and engineering*. Perseus Books, New York.

Thattai, M and van Oudenaarden, A. (2001). Intrinsic noise in gene regulatory networks. *Proc Natl Acad Sci U S A* 98(15): 8614–8619.

Turk, G. (1991). Generating textures on arbitrary surfaces using reaction-diffusion. *Siggraph 91 Conf Proc* 25: 289–298.

Zheng, Y and Sriram, G. (2010). Mathematical modeling: bridging the gap between concept and realization in synthetic biology. *J Biomed Biotechnol*. Available at: http://downloads.hindawi.com/journals/jbb/2010/541609.pdf [Accessed 7 November 2011].

Zi, Z, Zheng, Y, Rundell, AE, *et al.* (2008). SBML-SAT: a systems biology markup language (SBML) based sensitivity analysis tool. *Bioinformatics* 9: 342. Available at: http://www.biomedcentral.com/content/pdf/1471-2105-9-342.pdf [Accessed 7 November 2011].

CHAPTER 7

Applications of Designed Biological Systems

7.1 Introduction

Biological systems surround and pervade every aspect of human existence. It is no surprise that the use of synthetic biology to design living systems will find numerous application areas with significant impact on society. In this chapter we highlight current areas of research on real-world applications, and suggest future areas of investigation.

The use of biological technologies for useful purposes dates back to the emergence of human civilisation itself. For example, humans have used yeast to ferment beer and wine for thousands of years. Today biological technologies are found throughout diverse areas of industry. Synthetic biology promises to accelerate the development of optimised and novel biotechnologies that will be integrated in many aspects of industry and life.

7.2 The Potential for Developing Applications of Synthetic Biology

It is often stated that the whole purpose of synthetic biology is to produce applications or products. A recent report (BCCC research BIO066A, published in June 2009) on the commercial potential of synthetic biology suggests that the expected market for synthetic biology in 2013 will exceed \$2.4 billion. Of this, the chemicals and energy sector is the largest share with a predicted market value of \$1.6 billion, followed by the biotechnology and pharmaceuticals sector with a market share of \$0.6 billion. Growth rates are predicted to be 81% and 49% in these areas, respectively. For this reason a number of major industrial corporations are keenly interested in synthetic biology and what it has to offer.

There are also a host of other reasons for the growing interest in synthetic biology. Because devices can be designed for a specific task or application, unlike naturally occurring organisms that may have many functions, they should, as a technology, function more reliably than non-designed systems and can be optimised for the desired application. In addition, the use and reuse of a set of parts means the cost of characterising sub-assemblies can be shared over a number of products, and will thus be lower than dedicated organisms. Finally, interfaces on both the output and input side can be designed to integrate with existing systems, such as optical sensors and CCD cameras, thus permitting conventional instrumentation to be used rather than biological assays.

7.3 Criteria for Developing Applications in Synthetic Biology

Synthetic biology devices are not suitable for all applications, and it is possible to define some criteria for areas where they will be most suitable for commercial development. First, the application area must be sufficiently homogenous to permit one design to work in a wide range of situations. Thus, for example, chemical synthesis is a suitable area, whilst personalised medicine is not because the latter would require a bespoke device for each person. Second, the application area must have a sufficiently large volume to justify the cost of development. Third, the environment that the synthetic system is used in must be compatible with the chassis in which it has been implemented. This is true of most biologically-based processes and should be considered as one of the initial specifications in the design. For example, since many reactions are exothermic, this has led to interest in thermophilic organisms that can withstand the higher temperatures found in fermentation and other process chambers, which is considered preferable to the other option of actively cooling the chamber, since cooling is costly, inefficient and slows down reaction rates. Fourth, the intended application should ideally be constrained to avoid contamination of the synthetic system by external agents, or migration of the system outside the planned sphere of influence. For this reason, industrial biotechnology applications, which constrain the use of the synthetic biological system to inside a reaction vessel, are ideal. Other methods of constraint include adhesion to a surface, particularly suitable for sensor systems. The danger of accidental release of the synthetic system into unintended areas can be minimised through the use of 'kill' or fail-safe switches and/or the use of non-living, abiotic chassis which are unable to grow and replicate.

A final consideration in selecting suitable applications for the commercialisation of synthetic biology is public acceptance or at least the absence of public opposition. Some of the factors surrounding this are covered in Chapter 9 and include considerations of public benefit, potential risk and likelihood of direct exposure to the synthetic biological system.

7.4 Challenges in Developing Applications in Synthetic Biology

Even if an application is suitable for the use of synthetic biology devices, there are a number of challenges to overcome before it can be implemented. These challenges include competition from existing processes which will be associated with significant pressures to remain in place in the form of previous investment in development and infrastructure. For this reason, synthetic biology approaches to pharmaceutical synthesis will have a harder time replacing existing synthesis routes, however inefficient these may be, as there is not the cost pressure to change. On the other hand, applications in the synthesis of new chemical entities, areas such as biosensors, where there are not chemical or electronic sensors readily available, or in biofuels, where the volumes are such that small improvements in efficiency can have significant cost advantages, will provide a ready area for synthetic biology applications.

Another challenge to be met is the need for robust and efficient scale-up. Most synthetic biology systems are initially developed in a laboratory environment using basic equipment such as shake flasks or 96 well plates at volumes of the order of a few litres or less. These processes must then be transferred first to bench scale reactors and then to pilot plants and industrial scale production, possibly up to tens of thousands of litres per day. Whilst this process occurs regularly in chemical engineering, it is by no means an easy one, often requiring the re-optimisation of experimental conditions at different scales. The cost of this scale-up is part of the investment required to commercialise synthetic biology applications. Fortunately, there are successful examples of how this can be achieved. For example, the biotechnology company Amyris has successfully brought processes for the production of artemisinin and biofuels to an industrial scale. It is important to note that the synthetic biology product must be made in sufficient quantity to meet the whole market, otherwise the market will not be able to rely on the product and will use non-synthetic biology alternatives, eventually eroding the market for the synthetic biology product.

One very pertinent current challenge is the need for validation. In the early years, while synthetic biology is being developed, there is an absence of reliable characterisation data for parts and sub-assemblies, meaning that in early applications, developers will need to perform their own characterisation and absorb the cost that entails. As the field progresses, however, the development of an infrastructure of datasheets for biological parts and chassis will accelerate the commercial development of synthetic systems as lower cost re-use of parts and their associated validation data will be possible.

The presence of a distribution network to transport the product from the factory to the user can also be considered as a potential hurdle. If the synthetic biology product is a living organism, then the environmental constraints for transport and storage will be more demanding than for electronic or chemical products, although not dissimilar to those of packaged sterile medical devices. If the product is a chemical produced by synthetic organisms in the factory, then the compatibility of the synthetic biology-produced chemical with existing distribution networks will aid its substitution. This is a big consideration for potential products such as biofuels, where there is a significant premium for the synthetic biology product to be compatible with existing hydrocarbon fuel stocks. In addition, if the product for distribution is a living synthetic organism, then it must be recognised that public attitudes to risk may demand a greater degree of control over the organism. To this end, a number of groups have introduced parts that stop cell reproduction or can be triggered to cause cell destruction. In summary, activities to overcome these challenges are expected to be a significant element of the continuing development of synthetic biology products and applications in future years. The remainder of this chapter gives details of a number of potential future applications of synthetic biology.

7.5 Constructing Microbial Cell Factories

Living systems, from single-celled microbes to complex multicellular organisms, are able to transform simple compounds and nutrients into complex chemicals, materials and structures from the nanoscale to the macroscale. Biological engineers have only just begun to harness natural diversity to produce valuable compounds. The concept of a microbial cell as a 'factory' is a useful concept to encapsulate ideas around engineering life as a production platform. For example, much like a factory, cells are able to take material and energy inputs (for example, glucose) and transform them in a step-wise manner to higher value materials. In this analogy enzymes represent the machines in the factory that process and assemble raw materials into finished goods. Microbial cells have been used as factories to produce a wide spectrum of value-added products.

A number of valuable compounds, from small molecules to proteins, are produced by organisms found in nature. However, these are often made by organisms that are not amenable to large-scale industrial growth and production, and are often produced at low yields. To circumvent these issues, bioengineers will express metabolic pathways or proteins in a heterologous host. These hosts include the baker's yeast *Saccharomyces cerevisiae*, the model gram negative bacterium *Escherichia coli* and model mammalian cell lines such as Chinese hamster ovary cells. These production hosts are able to grow rapidly and to high density in fermenters and are genetically tractable. In addition to producing naturally occurring products in heterologous organisms, synthetic biologists are designing cells to produce chemicals, materials and proteins with no natural analogues. This is an example of using biological systems to produce completely synthetic compounds.

7.5.1 *Protein products*

One of the first uses of microbial cells as production platforms was the heterologous production of insulin in *E. coli* in 1977. The vast majority of human insulin on the market is now produced by bacteria. Protein-based therapeutics (known as biologics in the pharmaceutical industry) is a rapidly growing area of the industry's portfolio of products that have been approved for the treatment of cancers, inflammatory diseases and cardiovascular disease. In addition to therapeutic proteins, the production of enzymes is an important focus of several industries. These industrial enzymes include proteases, amylases and lipases for food and animal feed processing.

A key step in biologics and protein manufacturing is the production of full-length, functional proteins with high yields in a controlled fermentation process. Current research in synthetic biology aims to design 'smart' production organisms that are tailored specifically for protein overproduction. The incorporation of synthetic genetic circuits to sense fermenter conditions and control cell growth and metabolism may have significant impact on biologics and enzyme manufacturing. For example, genetically encoded sensors of dissolved oxygen or pH could control the expression of genes involved in central metabolism, rerouting carbon flow to avoid excessive acidification of the process.

7.5.2 *Fuels*

The derivation of liquid transportation fuels from petroleum is one of the most pressing challenges society faces today. Concerns over climate change, limited supplies and the geopolitical costs of oil extraction have driven the search for technologies to transform renewable biomass into fuels. Many organisms, notably baker's yeast, are able to ferment sugars into ethanol and do so with high productivities. The fermentation of corn sugars to ethanol is currently used to supplement gasoline as an oxygenate. However, ethanol has a relatively low energy density and is hygroscopic (takes up water), which presents problems in its distribution and use.

Synthetic biology offers the chance to reprogramme organisms to make unnatural or modified compounds, which have more favourable fuel characteristics. A recent example of creating synthetic metabolic pathways for the production of liquid fuels is the engineering of branched chain alcohol pathways in *E. coli* and yeast. Alcohols such as propanol and butanol have energy densities similar to petroleum-derived gasoline. This synthetic pathway uses two heterologous enzymes to divert precursors from amino acid biosynthesis into branched alcohols. Several other synthetic metabolic pathways have been demonstrated that produce components of diesel fuel (long chain alkanes and alkenes).

7.5.3 *Commodity chemicals*

In addition to liquid transportation fuels, there is a demand for renewable sources of commodity chemicals. Commodity chemicals are broadly defined as low molecular weight precursors of high value industrial chemicals. These are also known as 'platform chemicals', because they provide a common set of building blocks that the chemical industry can use to produce a diverse range of valuable products. Examples include ethylene, propylene and butadiene, which are produced in millions of tons per year to synthesise plastics, resins and other materials.

There have been several synthetic approaches to producing commodity chemicals from renewable biomass. DuPont has commercialised a genetically modified *E. coli* that produces 1,3-propanediol (1,3-PDO) from corn syrup. 1,3-PDO is used to make polyesters, composites, laminates and coatings. The biological process uses a renewable carbon source and uses 40% less energy than the conventional process. Synthetic biological routes to valuable compounds often have advantages in process energy costs, due to the fact that biological systems (enzymes and metabolic processes) operate close to room temperature and at standard pressure.

7.5.4 *Materials*

Many protein-based biological materials have physical properties that exceed even the most high performance man-made polymers. Some of the most well-known examples are spider silks. Spider silks are made from proteins containing highly repetitive proline-rich helical domains. Silks are able to stretch to many times their length without failure and have tensile strengths exceeding that of Kevlar®. Due to the difficulties in farming and harvesting silk protein from spiders, there is significant interest in producing silks in microbial cells. This presents particular challenges: highly repetitive proteins often express poorly in recombinant hosts due to inefficient codon usage. To solve this, several researchers have used chemical gene synthesis to construct synthetic spider silk genes for expression in bacteria. These genes have identical amino acid sequences to native spider silks, but are computationally codon-optimised for production in fast-growing microbes such as *E. coli*. The 'recoding' and chemical synthesis of genes is a way to eliminate the need to extract DNA from a source organism and is a strategy to circumvent difficulties in heterologous expression as well.

7.5.5 *Specialty chemicals and drugs*

Living systems produce a number of highly complex chemical structures, especially through secondary metabolism. For example, plants produce a wide range of structures with multiple chiral centres that are involved in chemical defence of the organism. Many of these natural products are bioactive in humans or are able to kill microorganisms and thus have therapeutic value. In fact, over half of all pharmaceuticals on the market are based on natural compounds. It is no wonder that synthetic biologists would seek to construct organisms to overproduce these valuable and often life-saving products.

Several groups have engineered microbes to produce fine chemicals for therapeutic purposes. *Saccharomyces cerevisiae* has been engineered to produce artemisinic acid, a precursor to artemisinin, a potent anti-malarial naturally found in the plant *Artemisia annua*. Currently, the chemical synthesis of artemisinin is cost prohibitive to the population of the Third World, where it is needed most. The researchers were able to use an engineered mevalonate pathway, an amorphadiene synthase and a cytochrome P450 monooxygenase from *A. annua* to produce artemisinic acid. The engineered yeast produced higher artemisinin yields than *A. annua* itself, although industrial scale-up and optimisation was required to make this route to production cost effective (see above).

7.6 Medical and Health Applications

One of the primary reasons for society to become proficient in engineering biological systems is for enhancing human health and well-being. The healthcare industry is one of the largest markets in the world, and directly impacts quality of life. As discussed above, synthetic biologists are working to engineer cells as factories to produce affordable drugs, protein therapeutics and enhanced materials. There are significant efforts in interfacing synthetic biological systems directly with patients to improve diagnosis, prevention and treatment of diseases and disorders.

7.6.1 *Biosensors*

The molecular and cellular processes in living organisms rely on the specific recognition of molecules, ranging from small metabolites to nucleic acids, lipids and proteins. This recognition can be exploited to sense, detect and measure biomolecules in a research, clinical or field setting and has been heavily exploited by the biotechnology industry. For example, antibody-based assays for biomolecules (such as the ELISA assay) are widespread and commercially available.

Synthetic biologists aim to make the design of biosensing materials and organisms rapid and amenable to forward design. Analytical devices capable of measuring biomolecular species in patients are invaluable in characterising normal and pathological processes and to alert clinicians to appropriate medical treatment. The most widespread and developed example is the blood glucose sensors, which are used by millions of people daily. However, these technologies are limited to the detection of one compound (glucose) and cannot be tailored to sense other biomolecules. The generation of biosensors remains very much a research project, where significant amounts of resources are required to develop unique solutions for each compound one desires to detect. Given the explosion in biomolecular data from laboratory and clinical studies, the lack of a platform for tailoring biomolecular detection is a fundamental bottleneck to the era of personalised medicine. A detection platform that could be rapidly tailored for nucleic acids or proteins would enable clinicians to take advantage of the wealth of genomic and proteomic data available.

7.6.2 *Smart therapeutics*

In addition to engineering microbes to produce therapeutics, several groups are exploring the use of live cells as therapeutics. Microbes in their natural state are endowed with many functions that could be utilised to discriminate between healthy and disease states (such as receptors and environmental sensing components) and act in a therapeutic manner (such as synthesising therapeutic proteins, invading diseased cells or synthesising chemicals, as above). Towards these aims, *E. coli* has been engineered to sense and destroy cancer cells by environment-dependent control of invasion. The basis of the system was invasin from the pathogen *Yersinia pseudotuberculosis* as an output that allowed *E. coli* to invade mammalian cells. To render the bacteria cancer cell-specific, invasin expression was controlled by several heterologous sensors: the *Vibrio fisheri* quorum-sensing circuit, the hypoxia responsive promoter or the arabinose- responsive promoter. Each of these is designed to induce invasin expression and bacterial invasion only in the presence of tumour cell environments (for example, tumours are highly hypoxic) or via external, researcher-inducible control. These engineered bacteria were shown to successfully invade several cancer-derived cell lines, demonstrating that cells can be programmed with sensors and outputs to achieve therapeutic functionality. Taken together, this work shows that natural functions of bacteria can be re-engineered and augmented to construct useful functions.

7.6.3 *Tissue engineering and patterning*

The coordinated organisation of cells in specific patterns is a classic example of complex function in many organisms and is central to the development of multicellular organisms from a single fertilised oocyte. Pattern formation typically involves signalling and communication between cells, processing of these signals and modulation of the expression of a variety of genes. The ability to design pattern formation will hold great utility in applications such as tissue engineering and biomaterials. Towards these aims, several groups have explored how collections of cells can be programmed to form user-specified patterns.

'Traditional' tissue engineering has focused on the development of three-dimensional extracellular scaffolds to direct the differentiation and patterning of cells into functioning tissues. A synthetic biology approach would also include the 'bottom-up' patterning of cell fate and position. In one example, researchers enabled *E. coli* cells to communicate with each other using the quorum-sensing system from *Vibrio cholerae*. Engineered 'receiver cells' were designed to express fluorescent proteins based on the concentration of the signalling molecule synthesised by a 'sender' cell. The receiver cells were designed such that they were responsive only to a defined range of signalling molecules, analogous to a bandpass filter. The range of signalling molecules the receiver cells was responsive to was tuned by changing the kinetic parameters of the underlying information processing circuit. Combinations of sender and receiver cells can be used to create two-dimensional patterns on a lawn of cells such as a bullseye, ellipses and clovers. Thus, the engineering of underlying functionality (i.e. quorum sensing) and gene circuits allowed spontaneous pattern formation in a population of cells.

Other approaches have been used to programme pattern formations that are inspired more by lithographic and printing techniques than developmental pattern formation. In a stunning example of engineering synthetic functions into organisms, a strain of bacteria that can sense red light and control gene expression was constructed. To accomplish this, the researchers constructed a chimeric two-component photorhodopsin system from the cyanobacteria *Synechocystis* in *E. coli*. When coupled to the expression of a chemical output, this function allowed a lawn of bacteria to act as a photographic film — projection of an image onto the lawn results in the recording of a high-definition two-dimensional chemical image at resolutions up to 100 megapixels per square inch. The control of pattern formation in living cells will have important applications in constructing complex patterned biomaterials, tissue engineering and parallel biological computation. In an extension of this work, synthetic biologists have enabled massively parallel 'edge detection' of a projected image such that cells communicate to discriminate and delineate boundaries between cells sensing light and dark regions of the image. These efforts hold great potential to explore how biology uses large numbers of computational elements (in this case cells) to compute complex problems and to combine 'top-down' lithographic-style patterning with 'bottom-up' parallel computation to specify patterns. This work was initiated as part of the International Genetically Engineered Machine (iGEM) student competition and is discussed further in Chapter 8.

7.7 Synthetic Biology for a Sustainable World

Living systems have evolved to gather resources from their environment for survival and replication. In doing so, they often drastically change the local or global environment. The ability of organisms to alter the environment and act 'beyond the bioreactor' is a frontier area of research in synthetic biology, but holds immense promise for tackling problems in sustainability.

7.7.1 *Bioremediation*

Bioremediation is the use of natural and engineered organisms to remove pollution from contaminated environments. This is usually accomplished by microbes that are able to mineralise (oxidise completely to CO_2) contaminating compounds. Although a number of bacteria have been identified that can degrade a wide variety of toxins and xenobiotics, *Pseudomonas* species have been the most intensively studied. One of the major bottlenecks to widespread adoption of bioremediation is a lack of understanding of microbial physiology and microbial community diversity in the field. Because of this, synthetic organisms engineered for purpose may be a useful approach to remediation. A future vision of bioremediation is to develop synthetic organisms that could persist in the environment, monitor their surroundings for toxins and pollutants and execute a programme of remedial steps. These synthetic organisms would act as persistent 'sentinels' in the environment but would clearly need to be considered in terms of public acceptance.

7.7.2 *Biomining*

The extraction of metals from ores is a major economic activity in virtually all parts of the world. The involvement of microbes in metal extraction (known as biomining) has a long history, although early miners likely did not know microbes were participating in the liberation of metals from ores. In modern biomining, acidophilic bacteria are used to solubilise metal sulphides or metal oxides by providing sulphuric acid, or to provide a pretreatment to break down the structure of ores such that chemical leaching can take place. So far, there have been few efforts in engineering biomining organisms for enhanced metal leaching, making this an area ripe for synthetic biology approaches.

7.7.3 *Engineering crops and commensal soil organisms*

One of the major problems facing society today is the availability of food, especially given projections of a rapidly growing global population, climate change and destruction of arable land worldwide. One of the frontiers of synthetic biology is the design and engineering of crops able to survive drought, high salinity soil and heavy metal exposure while producing high yields of safe produce. This vision would be a dramatic extension of the technology in genetically modified crops found today. Another possible route is the engineering of microbial commensal organisms in the soil of farmland to increase nutrient availability and cycling. The development of these types of technologies will have to be closely integrated with local community needs within a regulatory framework.

Reading

Alper, H and Stephanopoulos, G. (2009). Engineering for biofuels: exploiting innate microbial capacity or importing biosynthetic potential? *Nat Rev Microbiol* 7(10): 715–723.

Anderson, JC, Clarke, EJ, Arkin, AP, *et al.* (2006). Environmentally controlled invasion of cancer cells by engineered bacteria. *J Mol Biol* 355(4): 619–627.

Bayer, TS. (2010). Transforming biosynthesis into an information science. *Nat Chem Biol* 6(12): 859–861.

BCC Research: Synthetic Biology: Emerging Global Markets. Available at: http://www.bccresearch.com/report/BIO066A.html [Accessed 6 September 2011].

Chang, MCY, Eachus, RA, Trieu, W, *et al.* (2007). Engineering *Escherichia coli* for production of functionalized terpenoids using plant P450s. *Nat Chem Biol* 3(5): 274–277.

Chang, MCY and Keasling, JD. (2006). Production of isoprenoid pharmaceuticals by engineered microbes. *Nat Chem Biol* 2(12): 674–681.

Ferrer-Miralles, N, Domingo-Espin, J, Corchero, JL, *et al.* (2009). Microbial factories for recombinant pharmaceuticals. *Microb Cell Fact* 8(17). Available at: http://www.microbialcellfactories.com/content/pdf/1475-2859-8-17.pdf [Accessed 7 November 2011].

Keasling, JD. (2010). Manufacturing molecules through metabolic engineering. *Science* 330: 1355–1358.

Lee, SK, Chou, H, Ham, TS, *et al.* (2008). Metabolic engineering of microorganisms for biofuels production: from bugs to synthetic biology to fuels. *Curr Opin Biotechnol* 19(6): 556–563.

Levskaya, A, Weiner, OD, Lim, WA, *et al.* (2009). Spatiotemporal control of cell signalling using a light-switchable protein interaction. *Nature* 461: 997–1001.

Prather, KLJ and Martin, CH. (2008). De novo biosynthetic pathways: rational design of microbial chemical factories. *Curr Opin Biotechnol* 19(5): 468–474.

Ro, DK, Paradise, EM, Ouellet, M, *et al.* (2006). Production of the antimalarial drug precursor artemisinic acid in engineered yeast. *Nature* 440: 940–943.

Rude, MA and Schirmer, A. (2009). New microbial fuels: a biotech perspective. *Curr Opin Microbiol* 12(3): 274–281.

Steen, EJ, Kang, YS, Bokinsky, G, *et al.* (2010). Microbial production of fatty-acid-derived fuels and chemicals from plant biomass. *Nature* 463: 559–U182.

Tabor, JJ, Levskaya, A and Voigt, CA. (2011). Multichromatic control of gene expression in *Escherichia coli*. *J Mol Biol* 405(2): 315–324.

Topp, S and Gallivan, JP. (2007). Guiding bacteria with small molecules and RNA. *J Am Chem Soc* 129(21): 6807–6811.

van der Meer, JR and Belkin, S. (2010). Where microbiology meets microengineering: design and applications of reporter bacteria. *Nat Rev Microbiol* 8(7): 511–522.

Verpoorte, R, van der Heijden, R and Memelink, J. (2000). Engineering the plant cell factory for secondary metabolite production. *Transgenic Res* 9(4–5): 323–343.

Widmaier, DM, Tullman-Ercek, D, Mirsky, EA, *et al.* (2009). Engineering the Salmonella type III secretion system to export spider silk monomers. *Mol Syst Biol* 5. Available at: http://www.nature.com/msb/journal/v5/n1/pdf/msb200962.pdf [Accessed 5 November 2011].

CHAPTER 8

iGEM

8.1 Introduction

The International Genetically Engineered Machine (iGEM) competition has played a major role in the development of synthetic biology and highlights how the subject is accessible to undergraduates and even high school students. While iGEM is primarily a teaching tool, training undergraduates in lab project work and the principles of synthetic biology, it has also produced many significant advances in synthetic biology, with many iGEM projects going on to be published as important scientific advances. In this chapter there are descriptions of iGEM projects from the first six years of the competition that are good examples of successful and interesting synthetic biology projects.

8.2 The iGEM Competition

The iGEM competition emerged out of project work organised at MIT by Tom Knight and Drew Endy during MIT's independent activity period (IAP) in 2003 and 2004. The IAP is a special four-week term run at MIT in January each year where students pursue creative and individual projects rather than typical study. Many IAP projects at MIT in the last few decades have been held as competitions and previous student competitions in coding and robot-building were the inspiration for iGEM.

In the 2003 IAP, synthetic biology students worked with the synthetic 'repressilator' system constructed previously by Michael Elowitz. Teams of students worked on improving and modifying this system following engineering principles. The 2004 IAP focused instead on spatial design projects, particularly those of pattern formation, and it stuck with the same standards of the year before, using and generating BioBricks™ parts, cataloguing these in a registry and documenting work online through the science wiki website, openwetware.org.

Through a one-off funding from the National Science Foundation of the USA, the IAP competition was opened up to four other American universities as a synthetic biology summer boot camp from June to November 2004, and culminated in a Jamboree at MIT where the teams from across the USA compared their ideas and results. The 2004 summer boot camp and Jamboree were so successful that they were repeated in 2005 as the first official iGEM competition, with 13 teams participating, including 3 from outside the USA.

After 13 teams in 2005, 2006 saw 32 teams from around the world competing and a more formal competition was established that included the awarding of a Grand Prize at the MIT Jamboree — an oversized aluminium brick. The competition was covered by international press and the unexpected winners were a very impressive team of students from Slovenia. From 2006 to 2010, the competition grew from 32 student teams to over 125, making it difficult to organise and judge the Jamboree, especially with over 1,500 people at MIT for the event. To encourage expansion and further participation, in 2011 iGEM introduced a two-stage competition with regional Jamborees in Europe, North America and Asia followed by a finalists-only Jamboree at MIT. With over 160 teams from every continent now participating, including teams from high schools and the armed forces, the iGEM competition is providing a hands-on introduction to synthetic biology for thousands of students each year.

8.3 How iGEM Works

The iGEM competition is for teams of high-school students, undergraduates or masters students with most teams being made up of between 5 and 15 students. Established researchers and PhD students usually supervise a team, putting them together in the first half of the year before the team works full time on their project through summer. By registering with iGEM, each team receives a 'kit' of around a thousand standard biological DNA parts in BioBrick™ form from the iGEM Registry that they can store in the freezer and use to help them build their project. These are parts that have been built, characterised and submitted by teams from previous years and include systems for light detection and coloured pigment production. Most parts are for work in *E. coli*, but some *Bacillus* and yeast parts are also distributed.

Registration for iGEM also gets each team space on the iGEM website to start a wiki, which gives them a website in which to document their project, its progress and the results. Initially these websites were part of the openwetware.org website, but are now fixed within iGEM's own website. The idea of wikis is to encourage open science, allowing others to see your real work and what you are doing in order to improve and build upon the science as quickly as possible. It also encourages teams to share resources, so that as many teams as possible can reach their goals during the short time frame of the project. It is also a crucial point of iGEM that the DNA sequence of new biological parts that students make each year is uploaded to the official parts registry website for use by the teams of the following years. This also makes these parts free to use for the synthetic biology community, so occasionally parts that are patent-protected are not uploaded. Importantly, iGEM asks that uploaded parts are well described and the results of any testing are displayed too. This improves future use of parts and advances synthetic biology every year.

Registration fees, laboratory work, DNA synthesis of parts and travel to the Jamborees all add up to quite a cost, so many teams engage with bioscience companies, government initiatives and local enterprises to get sponsorship for their teams. iGEM projects normally involve a considerable amount of outreach, with students sometimes teaming up with artists, designers, film-makers, medical doctors, charitable foundations, lawyers, sociologists and even NASA and the FBI in order to further expand their projects in interesting directions.

By the end of summer each team will produce a wiki of their project and get themselves team T-shirts. For the Jamboree they will make a poster to display and give a presentation of their work. At the Jamboree there are different 'tracks' where projects that fall into specific themes like software development, healthcare and energy compete against one another. An army of established members of the synthetic biology community judge each team on a variety of different criteria so that on the final day of the Jamboree awards can be given. Teams are awarded bronze, silver and gold medals for fulfilling basic criteria set out each year and then individual prizes are given for the best projects in each track. On top of this, prizes are awarded for projects that stand out in key aspects, such as the best mathematical models, the best wiki, the best poster, the best human practices and the best new part. The final overall winning team and the runners-up are usually teams that are strong in all aspects, engaging through outreach, presenting well but also producing real results on interesting new synthetic biology ideas.

Over the course of an iGEM project students will work very hard in both the laboratory and on modelling, presenting and promoting their projects. But unlike many other courses that undergraduates can take in summer, iGEM gives them a chance to do real cutting-edge research science, introducing them to synthetic biology and open science while also allowing them to contribute to the new technologies and direction of the subject. Dozens of projects have gone on to later be published as major research articles in the top academic journals, including *Nature*, *Science* and *Cell* showing the quality of the research ideas iGEM has generated in synthetic biology. On top of this, many students choose to follow iGEM with synthetic biology research PhDs. Already, ex-iGEM students have now founded their own research groups and biotechnology companies in many places around the world.

8.4 Examples of iGEM Projects

8.4.1 *Bacterial Photofilm (UT-Austin and UCSF 2004)*

In 2004, the team from the University of Texas at Austin engineered *E. coli* to act as a photographic film, where projected images on a lawn of bacteria could be recorded by the production of a pigment. This approach enables biological pattern formation that is inspired by lithographic and printing techniques rather than the developmental pattern formation observed in nature, an example of truly synthetic biology.

To accomplish this, the UT-Austin team used a chimeric light-sensing system developed by the Voigt Lab at the University of California, San Francisco. This is a synthetic two-component sensing system, with a light-sensing protein domain taken from a cyanobacterium coupled to a response regulator from *E. coli*. The UT-Austin team put the gene lacZ under the control of the light-responsive system. LacZ is able to cleave a small molecule to produce a black pigment. Coupling these functions allowed a lawn of bacteria to act as a photographic film — projection of an image onto the lawn results in the recording of a high-definition two-dimensional chemical image at resolutions up to 100 megapixels per

Fig. 8.1 An example of a bacterial photofilm made using *E. coli* cells producing pigment in response to light activation. (Courtesy: BioBrick™ Foundation iGem Competition.)

square inch (Fig. 8.1). The results from this project were visually striking and even though the work was just an iGEM student project it gathered enough attention to be featured and written up in the prestigious international journal *Nature*.

The control of pattern formation in living cells will have important applications in constructing complex patterned biomaterials, tissue engineering and parallel biological computation. In an extension of the iGEM project work, Jeff Tabor and Chris Voigt went on to enable massively parallel edge detection of a projected image, where cells communicate to discriminate and delineate boundaries between cells sensing light and dark regions of the image. They connected additional synthetic device modules to the light-sensing system built by the 2004 iGEM project, with each module constructed independently, characterised and modelled to redirect the behaviour of the full system. In a research article publication of this work, Tabor *et al.* engineered a cell–cell communication component and a simple genetic logic gate such that cells in the dark areas of the image produce a diffusible chemical signal (see Tabor 2009 and Chapter 5). By implementation of genetic logic, only cells that sense the light areas of the image and the diffusible signal are able to generate a pigment as a read out of the image edge. These efforts hold great potential to explore how biology uses large numbers of computational elements (in this case, cells) to compute complex problems and to combine 'top-down' lithographic-style patterning with 'bottom-up' parallel computation to specify patterns.

8.4.2 *Eau d'e coli (MIT 2006)*

MIT not only host the iGEM competition but have also entered a team every year of the competition. In 2006, the MIT team won the award for best system, with a project that brought odour into the genetic engineering toolbox. Their project, *Eau d'e coli*, was a system engineered to give the production of different smells from *E. coli* bacteria at different points in a typical growth cycle. While the idea of bacteria smelling of bananas is fun, the project had a useful application; researchers often have to assay whether cells are in exponential growth or have moved to stationary phase and being able to distinguish the two different growth phases from smell would simplify this procedure.

MIT's team chose two strong scents; for exponential growth phase the cells would produce wintergreen oil (methyl salicylate) and when in stationary phase the cells would switch to producing a banana smell (isoamyl acetate). To achieve this requires adding and optimising all the genes for the enzymes involved in the biological production of these two chemicals and also regulating these genes so that they only express in the correct growth phase. The project was therefore both a metabolic engineering project and a gene regulation project. Wintergreen production required two enzymes to convert *E. coli* endogenous chorismate into methyl salicylate, and banana required three enzymes to convert endogenous L-leucine into isoamyl acetate. The genes for these enzymes were taken from diverse microbes and placed into an *E. coli* strain which was known not to have its own strong smell. To regulate the two pathways required a promoter part that was activated in stationary phase only and an inverter device that would shut down another promoter when stationary phase occurred.

The MIT team in 2006 were notable not just for appearing at the Jamboree with bacteria that really smelled strongly of bananas and wintergreen but also for breaking down their project into manageable sub-parts. The project was rigorously described in terms of parts that were linked into devices that then came together to make systems. This abstraction allowed different members of the team to work on their own section before these were brought together to make the final product. Each section was planned to have testable goals so that they could be handled independently before being combined. By following this engineering principle the students achieved a remarkable amount of their project in one summer, producing both a wintergreen scent-producing *E. coli* strain and a banana scent-producing strain. Their project was a big hit at the Jamboree and was later featured in magazines and on American National Public Radio.

8.4.3 *Towards self-differentiated bacterial assembly line (Peking 2007)*

Peking University were the winning team at the 2007 Jamboree with a project that designed and implemented two novel synthetic systems, both of which contribute to a larger goal of engineered *E. coli* cells that can be programmed to differentiate into subpopulations performing different tasks. Programmed differentiation like this is attractive as it could allow division of labour within a cell population.

The two systems Peking designed were a hop-count system for spatial differentiation and a push-on-push-off switch system for temporal differentiation. The hop-count system made use of the natural bacterial conjugation system for transfer of plasmids from one cell to a neighbour. By editing and rearranging specific DNA parts required in conjugation plasmids, Peking were able to develop a system where the plasmid being transferred lost

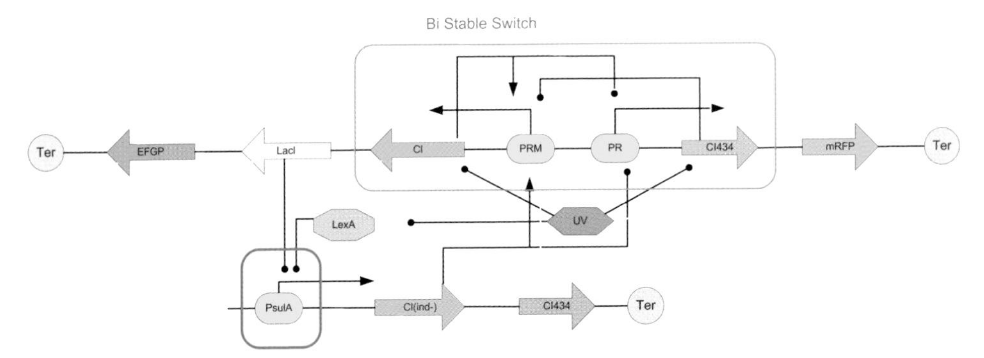

Fig. 8.2 The Peking 2007 iGEM team's circuit design for a UV-activated push-on-push-off switch constructed from a NOR gate interfaced with a bistable switch. (Courtesy: http://parts.mit.edu/igem07/index.php/Image:Peking_Swich_fig.1.JPG.)

specific sections of DNA each time a transfer occurred, effectively making it a counter device that went from one cell to the next.

The second system, the push-on-push-off switch, was based on the well-known bistable toggle switch system first described in 1999. To an ultraviolet (UV) light-controlled toggle switch, Peking added a NOR logic gate device (Fig. 8.2). The NOR gate compiles the information from both the input (UV light) and the memory of the toggle switch and uses this to flip the toggle switch when there is an induction, even though it is the same inducer each time. The design of this system is interesting as it nicely interfaces an existing synthetic system, the toggle switch, with a different device, the NOR gate, in order to now get a new system. The students built a mathematical model of this system to allow them to simulate its behaviour. They then went on to build and characterise a working bistable toggle switch as well as several NOR gates. Following the iGEM competition in 2007, work on this system continued and in 2010 the team published a major research article describing a working push-on-push-off switch and the many steps it took to create and improve it (see Lou 2010).

Peking's project gave iGEM many useful new parts for constructing new devices and systems in *E. coli* but also put together existing parts in a new and interesting way. Their project shows that basic biological parts already found in bacteria, such as conjugating plasmids and UV responsive genes, can be rewired in a predictable fashion to give new functions.

8.4.4 *Bacto-builders (BCCS-Bristol 2008)*

The Bristol Centre for Complexity Sciences iGEM team (BCCS-Bristol) won the Best Modelling Prize at the 2008 Jamboree. Their project aimed at programming a fleet of *E. coli* cells so as to force them to collaboratively perform a task which would be too great for any individual cell. More specifically, the task targeted by this project involved the physical movement of particles through direct contact with a swarm of bacteria working in a coordinated manner to displace it to a desired location. The motivation was to make collective behaviour emerge by forcing bacteria to adhere to a set of simple rules so that physical particles are assembled according to some desired pattern.

In their design, collaborative displacement of the physical particles was made possible through the use of a search, signal and collaborative push mechanism. The simple rules to which each bacteria needs to adhere are (i) to randomly search the environment for a suitable particle; (ii) upon contact with the particle, stick to the particle and attempt to chemotactically move it towards a goal (a chemoattractant gradient); (iii) while trying to move towards the goal, send out a short range quorum signal to inform nearby bacteria that a particle is in the vicinity; and (iv) upon sensing of the quorum-sensing signal, activate response to the chemoattractant goal and chemotactically move towards it. A design alternative for (ii) and (iv) that was also envisioned and tested *in silico* is to directly produce the chemoattractant instead of the quorum-sensing signal upon physical contact with the particle.

Using these simple rules, the assembly of conglomerates of physical particles in a specific configuration becomes possible by altering the patterns of the chemoattractant gradient. The implementation was based on the re-engineering of some of the natural mechanisms used by *E. coli*, i.e. chemotaxis, environmental sensing and cell–cell communication.

Due to the complexity of the different aspects involved in the project (such as chemotaxis, physical force exerted between large numbers of cells in an environment, chemical diffusion of the quorum-sensing small molecule, fluid dynamics and individual cell states estimation), the BCCS-Bristol team chose to adopt the engineering design cycle described in Chapters 2 and 6 with a combination of modelling, model-based computer simulations (both deterministic and stochastic) and experiments. Computer-based simulations were used to predict the behaviour/performance of the design alternatives and improve them, while experiments were used to test the feasibility of the different design options. In particular, two types of model were developed: an internal differential equations representation of the synthetic gene regulation circuits embedded in each single bacterium, and an external representation of the system as a whole using computerised stochastic agents. Both types of model were extensively simulated *in silico*. Matlab® software was used for tasks such as numerical integration and statistical analysis, while Java programming language was chosen for the development of a stochastic agent-based framework to simulate bacterial chemotaxis, the physical interactions between the bacteria and particles and the chemical fields exhibiting diffusion. The simulation of a large number of bacteria required huge amounts of computing power that were provided by the Blue Crystal high performance computing cluster at the University of Bristol. What sets this project apart and allowed the team to win the Best Modelling Prize is the use of a rigorous model-based approach for the design and analysis of their design options prior to experimental implementation.

8.4.5 *E. chromi (Cambridge 2009)*

The Grand Prize at the 2009 Jamboree was awarded to the University of Cambridge team for their project *E. chromi*. In this project, the student team set about producing and characterising a series of pigment-producing genes and a set of regulation elements known as 'sensitivity tuners'. Vivid colours are found everywhere in nature and the project sought to bring these to *E. coli* so that they can be used as reporter genes, in place of typically used reporters such as GFP. The motivation for the work was that biosensors produced by synthetic biology would ideally need to be low cost and simple to read from. Measuring GFP output from a bacterial sample requires fluorescence equipment that would not be realistic in a rural situation, whereas the growth of coloured bacteria would be easy to interpret simply by a human eye. The sensitivity tuner parts added to the project by offering a way to filter biosensor inputs (e.g. the detection of arsenic in water) so that the pigment is only produced in response to a meaningful level of detection. The students imagined their bacteria could be put into a dipstick used to detect heavy-metal pollutants. The dipstick would produce different colours depending on the metals present.

The Cambridge team took a range of approaches when generating the pigment-producing genes. For orange and red pigments they used production of the carotene and lycopene metabolites, respectively. These are produced in the carotenoid pathways, and BioBricks™ parts for the four required enzymes of this pathway already existed thanks to previous iGEM teams. For the black/brown pigment, a single MelA gene was taken originally from *Rhizobium etli*, which produced an enzyme that catalyses the production of the melanin pigment from tyrosine. This gene had already been described elsewhere for use in *E. coli* but the Cambridge team still had to use PCR techniques to convert this into a usable BioBrick™. Having used existing parts and used PCR to generate new parts, the Cambridge team then relied on direct DNA synthesis to make the last set of pigment-producing parts; synthesising a 5-gene operon of the violacein pathway from *Chromobacterium violaceum*, a naturally occurring purple bacteria. The 5-gene operon produced a purple pigment in *E. coli* but was also designed to be easily edited to 4 genes to instead produce green pigments.

A real success of the Cambridge project was that the team produced attractive and vivid results, showing *E. coli* growing in a spectrum of colours and producing colour pictures on agar plates (Fig. 8.3). While this captured the imagination of the audience at iGEM, the team also included extensive characterisation data for all their parts and constructs and made efforts to highlight potential real-world applications of the project. For this they had teamed up with two professional designers to brainstorm how the technologies they had developed in their project could impact on the future world. The design side of the project gave it extra

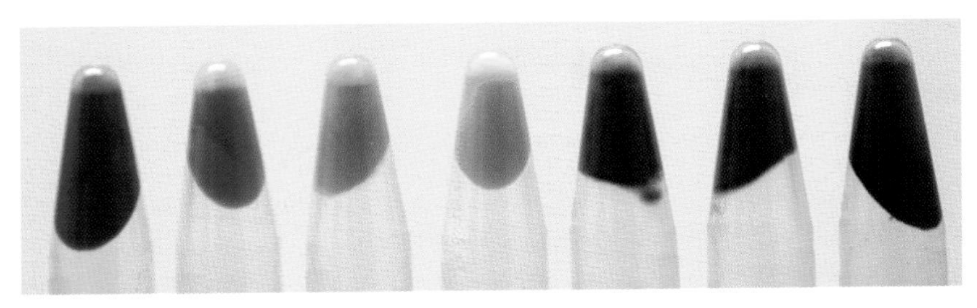

Fig. 8.3 Pigmented *E. coli* strains engineered by the 2009 Cambridge iGEM team. (Courtesy: BioBrick™ Foundation iGem Competition.)

impact and a conceptual art work, *E. chromi*, based on the engineered pigment-producing biosensor *E. coli*, has gone on to win nominations at prestigious art and design awards.

8.4.6 *Parasight (Imperial College London 2010)*

The Best Human Practices Prize at the 2010 Jamboree was awarded to Imperial College London's *Parasight* project. Human practices are the consideration of how a technology could potentially impact society and the wider world. This includes but is not limited to: considering the potential hazards involved with use and risk to the environment upon release; design of devices based on the requirements of the end user; and working to spread knowledge and awareness of synthetic biology to an increasing number of people. It has been suggested for a long time that it is these human practices that should either define or at least heavily influence the specifications for a biological device or system. The *Parasight* project was the first project at iGEM to take its specifications directly from human practices — what the end-user (aid agencies and possibly communities in the field) would require from the product, while ensuring safety and increasing awareness of the device itself and the technology behind it.

The aim of the project was to develop a modular method for sensing parasites in water (specifically *Schistosoma*) that would trigger a very rapid response that would be detectable without sophisticated equipment in the field, potentially by relatively untrained users. By consulting *Schistosoma* experts, the end-users of such a sensor were suggested to be parasite tracking teams and local communities attempting to determine which water sources were safe to use for activities such as bathing and washing clothes. To be useful to these groups of people, the sensor would need to be as cheap as possible with the possibility of use without sophisticated equipment (such as fluorescence detectors). In particular for use by communities, the result would need to be determined very rapidly rather than in the order of several hours like many pigment- or fluorescent-based systems. As an environment sensor, of paramount importance was safety as there would always be a risk of user exposure or environmental release.

The parasite-sensing device was designed to include a peptide signalling system that enables detection of the proteases that parasites are required to secrete in order to penetrate human skin. This sensor would connect to a fast, visible output module by production of the very small protease TEV. The output system was developed from an enzymatic tetramer (the XylE gene), which produces a yellow compound visible at low concentrations. A modified version of this enzyme was inactivated by fusion to another protein but could be released, by cleaving off this other protein using TEV protease (Fig. 8.4). Upon TEV expression, large amounts of enzyme are rapidly freed to generate a signal visible to the human eye in less than a minute. The system also doubles as a 'kill-switch', reducing risks of release, as the visible yellow product of the reaction is toxic to bacteria hosting the system.

To further increase the suitability for use in a real-world situation, the system was designed for use in the chassis organism *B. subtilis*, as most strains of this are already engineered to be unable to survive without at least one key nutrient (and therefore unlikely to survive in the wild). These bacteria also naturally form spores for easy transport and

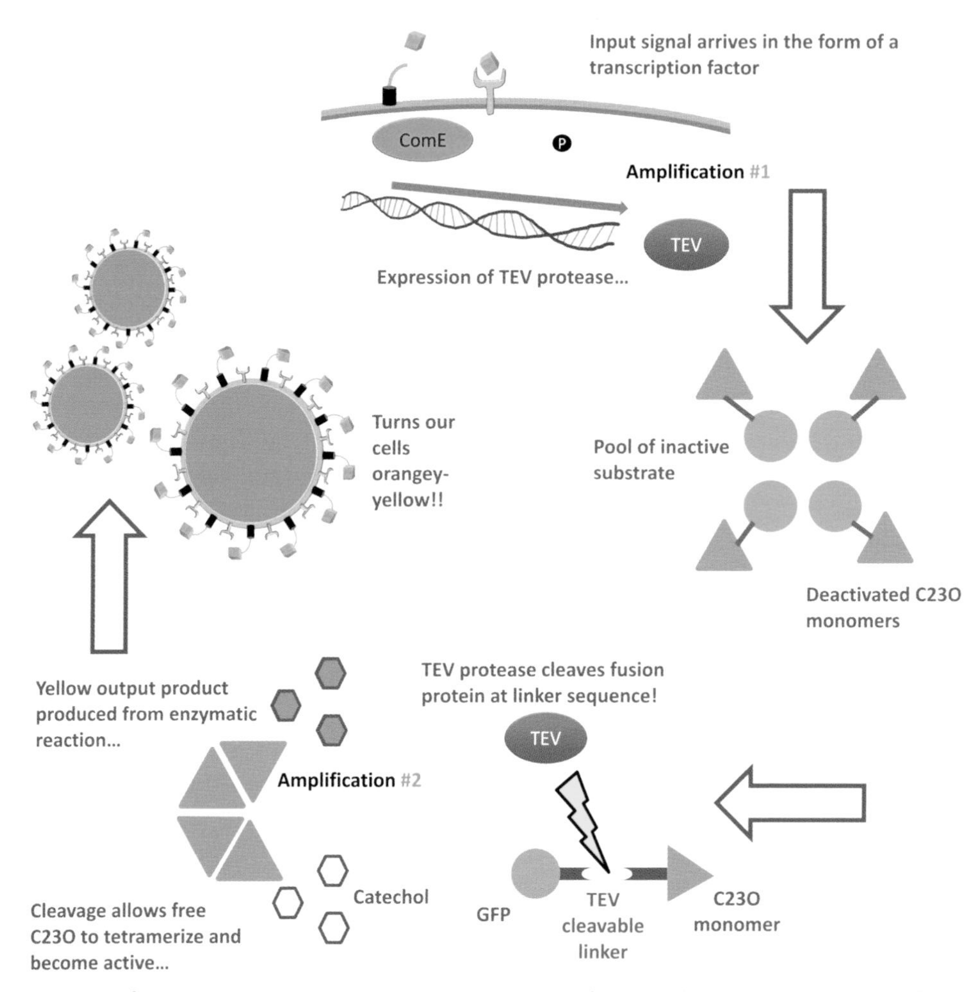

Fig. 8.4 A fast-response reporter module based on the gene XylE, from Imperial College London's 2010 iGEM project. (Courtesy: BioBrick™ Foundation iGem Competition.)

for growth in the field, and they readily carry out DNA integration and excision, allowing the DNA parts of the system to be stably inserted into the cell genome with no antibiotic resistance gene needed in the final product. In the *Parasight* project, the students also managed to design and produce prototype kits for their application that were made to attract *Schistosoma* in water while keeping the engineered bacteria out of the environment. A user manual was produced with these prototypes to spread awareness about the technology underpinning the sensor and its safe use.

While other biosensor projects have been seen in iGEM many times before, the *Parasight* project stands out because the design and specifications of the project were based on human practices from start to finish — from initial consultation with experts and individuals who work with parasites in the field, from early discussions with a safety and ethics panel and with potential end-users of the technology, through to later work raising awareness with visits to local schools and creating prototype user kits and manuals.

Reading

Learn About iGEM. Available at: http://ung.igem.org/IGEM/Learn_About [Accessed 6 September 2011].

Levaskya, A, Chevalier, A, Tabor, J, *et al.* (2005). Synthetic biology: engineering *Escherichia coli* to see light. *Nature* 438: 441–442.

Lou, C, Liu, X, Ni, M, *et al.* (2010). Synthesizing a novel genetic sequential logic circuit: a push-on-push-off switch. *Mol Syst Biol* 6: 350. Available at: http://www.nature.com/msb/journal/v6/n1/full/msb20102.html [Accessed 7 November 2011].

OpenWetWare. A wiki-style online community for bioscience. Available at: http://openwetware.org [Accessed 6 September 2011].

Registry of Standard Biological Parts. Available at: http://partsregistry.org/Main_Page [Accessed 6 September 2011].

Tabor, JJ, Salis, HM, Simpson, ZB, *et al.* (2009). A synthetic genetic edge detection program. *Cell* 137(7): 1272–1281.

CHAPTER 9

The Societal Impact of Synthetic Biology

9.1 Introduction

Is synthetic biology something radically new, or is it merely an extension of molecular biology? Many argue that, while it is the next big step in taking biotechnology forward, it builds on technologies for sequencing and manipulating DNA that have been developed since the 1970s. The last decade has, however, been a time of tremendous improvement in the ease of using many of these technologies, and this, along with rapidly falling costs and the dispersion of experimental approaches once thought to be the domain of elite biologists, has resulted in the dissemination of synthetic biology both widely, among sectors and academic approaches, and deeply, from Nobel Prize winners to high school students and amateur biologists.

Although not unique to synthetic biology, the combination of easy access to synthesised DNA, powerful computers to aid design, and the distribution of these technologies to users beyond the 'traditional biologist' suggests a radical departure from traditional biology, which raises unique safety and security concerns, as well as questions about ownership. Further, the idea of using these technologies to construct minimal cells and living organisms has sparked discussions about whether such ambitions are ethical and, beyond that, what 'creating life' means. The industrial application of synthetic biology, which would displace existing sources of some products and place ownership and control in the hands of large corporations, also raises concerns about the distribution of benefits of the technology and about global inequalities.

In this chapter we discuss five major areas of concern with respect to the societal impacts of synthetic biology. These include biosafety and the environment; biosecurity and biohacking; ownership; philosophical and theological issues; and public value and new global inequity.

9.2 Public Health and Environmental Risks

Synthetic biology revisits many of the safety concerns first identified with the development of recombinant DNA technology. Although we now have many years' experience of dealing with these issues, and have both statutes and regulatory guidance, most of those coming into synthetic biology today are young, and were not of the generation involved in the debates about recombinant DNA. It is therefore important to actively stimulate a generational transfer of knowledge about safety issues surrounding the sequencing and manipulation of DNA. Further complicating the picture, however, is that many of the people coming into synthetic biology today are not biologists. They are computer scientists, electrical engineers, mathematicians and physicists who have very little experience with microbiological safety. So in addition to gaining access and transmitting knowledge across a generational gap, there is also a need to transmit knowledge across cultural divides between scientific disciplines (Lentzos *et al.* 2008).

Of more concern is what happens when synthesised organisms leave the laboratory. Commentators highlight the risk that these organisms might behave in unexpected ways outside the lab, and have unintended and unanticipated detrimental effects on public health and on the environment. Many countries have experience in dealing with accidental releases of genetically modified organisms, but there is no prior experience of how truly novel organisms would act outside the laboratory environment. Although highly modified microorganisms are unlikely to survive in a natural environment, we need to consider the risks of synthetic organisms transferring their genes into existing organisms and altering the balance of the ecosystem. Synthetic organisms might also interact with naturally occurring substances leading to unexpected consequences, or they might evolve beyond their bespoke functionality and in that way elicit unanticipated side effects on the environment and other organisms.

With respect to deliberate or planned releases, synthesised organisms would be subject to the same regulations as any other genetically modified organism. The question is whether organisms should be subject to a greater degree of scrutiny because their genomes have been modified using synthetic DNA, rather than DNA extracted from another organism.

Key questions related to biosafety and the environment are outlined below:

Key messages

- Awareness of, and practices related to, biosafety need to be transmitted across generational and cultural gaps.
- Risks of unintended and unanticipated detrimental effects on human health and on the environment must be considered in synthesising organisms.

Key questions

- Who should be responsible for setting standards or determining the safety of synthetic organisms?
- Who should bear the burden or risk?
- How much should society spend on reducing risks from synthetic organisms to humans and the environment?

9.3 Biosecurity and Biohacking

In Chapters 7 and 8 we learned about some of the promises and possible applications of synthetic biology. Yet some of these beneficial applications could also be used in harmful or unintentionally dangerous ways. Concerns about this dual-use potential — where the knowledge, tools and techniques of synthetic biology, although holding great promise, may also be used to deliberately construct microorganisms that pose unknown dangers to public health and the environment — have been raised from the outset of the social and ethical debate on synthetic biology. In particular, some argue that synthetic biology's potential to enable an act of bioterrorism is a significant threat that must be carefully assessed and managed in light of ongoing advances in the field (CIA 2003; ETC Group 2007). Concerns have been raised as to synthetic biology's potential to 'deskill' the art of genetic engineering, allowing more people to genetically modify existing biological agents or create entirely new organisms with novel properties. Such agents could exhibit enhanced virulence, lethality, antibiotic resistance, greater environmental stability or perhaps even be tailored to induce changes in behaviour, cause chronic illness or be targeted to attack certain ethnic groups (Mukunda *et al.* 2009).

However, the nature of the synthetic biology threat is uncertain and these scenarios are contested; indeed there are those who argue that excessive emphasis on bioterrorism and biosecurity threatens to narrow the debate on synthetic biology and negatively impact the development of the science (POST 2008; Lentzos 2009; National Research Council 2009). It has also been noted that bioterrorism fears are more pronounced in the USA than in Europe, where concerns are less focused on biosecurity, and range across a variety of social and ethical issues, including biosafety and biosecurity considerations, but also unintended environmental consequences and intellectual property concerns. Nonetheless, on both sides of the Atlantic there has been a robust dialogue surrounding synthetic biology's potential for deliberate misuse, which has important implications for how the science is perceived and governed.

At the heart of the debate on synthetic biology's perceived security implications is the concern that it will permit individuals (who might otherwise be prevented due to natural or artificial barriers) to acquire dangerous pathogens (viruses or bacteria). Key to this debate is the assumption that synthetic biology will achieve its stated aims. That is, that it will become a true engineering discipline, offering improved means for doing biological engineering, enabling, among other things, the *de novo* synthesis of novel microorganisms that can perform any number of tasks, from ingesting greenhouse gases to producing biofuels. But although progress has most certainly been made (one need only consider the achievements of the student teams at the annual International Genetically Engineered Machine (iGEM) competition), doing biology remains a messy and difficult business, which is not *yet* amenable to amateur scientists constructing an Ebola virus in their garage.

There are, however, indicators (or perhaps 'warning signs'), which inspire equal measures of admiration and trepidation. Take, for example, the frequently quoted construction of poliovirus from commercially ordered gene sequences (Cello *et al.* 2002) or the resurrection of the 1918 Spanish Flu virus (Tumpey *et al.* 2005). These achievements are, indeed, proof of principle, demonstrating the capacity of synthetic genomics to (re)construct microorganisms that have caused, and could again cause, tremendous harm to human populations. Yet, as some have argued (see, for example, Vogel 2008), these research projects were precisely that, *research projects*, which required considerable skill, troubleshooting and tacit knowledge to achieve. Even for professional scientists, things do not always go as expected in the laboratory. In the case of Cello *et al.*'s synthetic poliovirus, insertion errors are believed to have greatly attenuated the infectivity of the virus (National Research Council 2004). Moreover, unlike poliovirus, many viral genomes (not to mention bacterial genomes) are not infectious on their own, but require 'booting' through the addition of other cellular components. For prospective bioweaponers, there are technical obstacles to converting a living pathogen into a viable biological weapon, which, contrary to popular belief, is a difficult process. For this reason, many experts argue that the present synthetic biology threat is, and will remain for the foreseeable future, largely limited to state-level programmes, as these have both the resources and the expertise to make use of synthetic biology for hostile purposes (Tucker and Zilinskas 2006; Lentzos 2009; Smith and Davison 2010).

Yet this is not to say synthetic biology will always be beyond the reach of terrorist groups, lone operators or, indeed, amateur biologists with malign intent. Some argue that such a scenario is possible, if not now, then in the future. As the knowledge, tools and techniques for doing synthetic biology proliferate, it will become, so the argument goes, increasingly easy for less skilled practitioners, hobbyists or do-it-yourself biologists to do biology, tinker with genomes and, perhaps, make novel living systems. The question behind all such predictions (whether implicit or explicitly stated) is: What will the future of synthetic biology look like? What if the tools of synthetic biology become widely accessible and easy to use? What if versatile error correction techniques are developed? What if desktop DNA synthesisers are in every home? As Robert Carlson (2009) has pointed out, rapidly declining DNA sequencing and synthesis costs are key drivers in pushing the limits of what can be done in synthetic biology, at least in terms of mapping increasingly complex genomes and synthesising ever larger gene sequences. There are also efforts to develop inexpensive tool kits and online protocols that can facilitate do-it-yourself biology (see, for example, the resources available on openwetware.org). Someday, if the trend continues, it might be possible for an amateur biologist to construct a gene of known function in his or her home or garage, and to insert it into an all-purpose chassis purchased from a local biotech supplier. Clearly, were a truly 'democratised' synthetic biology to emerge, it would not only enable any number of productive scientific endeavours, it would also pose dangers to public health and the environment, whether by accident or by design.

With a view to just such a possibility, some experts describe the potential emergence of a synthetic biology-enabled 'biohacker' culture, where amateur biologists test their skills and the practical limits of do-it-yourself biology to invent new, high performance, and possibly dangerous, microorganisms, much in the same way as contemporary computer hackers invent open source software, malware, etc. (Schmidt *et al.* 2008). Although less frequently discussed in the literature, in addition to so-called 'bioterror' and 'bioerror' concerns, one might also consider the unintended implications of 'biopranking', which, in keeping with the spirit of 'creativity' and 'play' often said to characterise the do-it-yourself biology community, might result in, for example, the intentional release of genetically modified bacteria into a neighbour's pool that, although intended only to turn the water green, could cause irreversible damage to nearby flora and fauna.

Thinking about such contingencies is valuable in the context of synthetic biology, as there are many unknowns that lie ahead and the expectations (for good and bad) are high. Therefore, to some extent, it is prudent to envision worst-case scenarios in an effort to be prepared for future catastrophes. Such thinking is not only endorsed by security analysts, but also by synthetic biologists (e.g. Church 2005). But there are also hazards in imagining and describing the future in a certain way, particularly when certain technological trajectories and the actions of certain social groups are presented as inevitable. The reason being, once spoken, there is a tendency to conceive of such scenarios as real, imminent and in need of action on the part of scientists, regulators, concerned interest groups and others. Thus, there is good reason to be cautious about particular renderings of danger and concern in the synthetic biology security debate, particularly those that are presented as 'almost certain'. It is important to ask why these particular threats and not others have been picked up on. It is also important to ask who should be responsible for assessing and managing the risks associated with synthetic biology at different stages of its development (NEST 2005; IRGC 2009). And, ultimately, what actions should be taken today to ensure the best possible future for synthetic biology and society?

These questions, and many more, belong in the public sphere, as they require the input of diverse stakeholders with contrasting interests and priorities. What is perhaps most challenging about the discussion of security as it relates to synthetic biology is that the number of worrying scenarios is potentially unlimited, as synthetic biology is a science that claims an open-ended capacity to design and construct novel living systems. For this reason, the values, views and concerns of scientists, academics, security experts and others are an essential element in imagining the future potential of synthetic biology and in governing its development. It is reassuring that the synthetic biology community has taken a proactive role in leading such a debate, and that initiatives like Europe's SYNBIOSAFE online conference (see Schmidt *et al.*'s 2008 summary report) have brought together participants from many countries and fields of expertise to discuss the issues of greatest interest and concern. Synthetic biology is a rapidly advancing field with enormous potential. No one can be certain what the future will look like. It is thus crucial to explore and debate as many perspectives as possible at an early stage of the development of the science, in order to develop the most robust ways of mitigating potential threats.

Key messages
• Synthetic biology is a dual-use technology, holding great promise while also posing unknown dangers to public health and the environment. • The potential for synthetic biology to enable an act of bioterrorism is a significant concern, but the nature of this threat is highly uncertain and contested. • Due to the inherent uncertainties associated with synthetic biology, there is a need for diverse stakeholders to discuss worst-case scenarios.
Key questions
• What security risks does synthetic biology enable today, and what might it enable in the future? • Who should be responsible for assessing and managing the risks associated with synthetic biology? • What actions should be taken today to ensure the best possible future for synthetic biology and society?

9.4 The Ownership of Technologies

Intellectual property (IP) rights recognise inventors and can help protect their creations as well as provide funds to finance the expensive commercialisation of ideas into products and applications. IP rights can, however, also significantly impede innovation. In synthetic biology, creators of new tools, techniques, chassis, biomechanical systems and even discrete BioBricks™ are wondering what kind of IP rights will best promote the field's development in a fair and responsible fashion.

Synthetic biology revolves around the complex intersection of engineering-driven application goals, computer-programmed design and biological-based materials, and as such couples software and biotechnology, both of which have had significant problems in their IP history.

Many synthetic biology practitioners draw a parallel between the sequences of DNA bases that they work with and the source code of software. This analogy works nicely, insofar as it allows synthetic biologists to explain how, ideally, the functional design in a biological machine is set within an alphabet (A-C-T-G) of encoded instructions, just as software programs are run by a set of numbers, letters and symbols, crafted by those versed in making this language translate meaningfully in electronics. However, as Kuman and Rai (2007) have noted, software has posed both copyright and patent difficulties as copyright protects works of expression that are not directly functional, while patents cover useful, novel and non-obvious inventions. Hence, software has run into the problem of being too functional to fit within copyright law, but also too algorithmic and non-tangible to be patentable. After a difficult early legal history, software has found at least some resolution. But IP concerns have also led to the development of open source alternatives, for example in the form of 'copyleft' licenses (Rai and Boyle 2007).

Synthetic biology is also intertwined with biotechnology's IP challenges. Patents are the predominant form of intellectual property in biotechnology and are granted for an invented product or process that provides a novel way of doing something or offers a new technical solution to a problem. Patents serve an important purpose in providing a legal framework to reward inventors and their co-patent applicants (often the funding bodies — universities, research councils, biotech industrial companies and venture capital firms, for instance) for translating ideas into products and applications. However, there have been many concerns, especially about the increasing uses of patents, which encompass quite basic biological pathways, which some believe have significantly limited access to and use of biological information, increasing transaction costs and thus inhibiting some important research endeavours (Hope 2008).

Synthetic biology faces analogous risks that progress will be stifled by a proliferation of both broad and narrow patents within biotechnology's broader IP culture. Several broad patents already exist in the field, held by universities and governments as well as more aggressive private firms. For example, there are foundational patents that cover the use of cellular machinery for information-processing tasks, mechanisms that modulate important cellular pathways and methods to select optimal DNA-binding proteins (Kumar and Rai 2007; Rai and Boyle 2007). When patents cover some of synthetic biology's most essential technologies, the prospect of inhibiting innovation is very real, as the cost of using patented information is often too high for early-stage researchers to undertake experimental work that might need to incorporate such technologies. In a similar vein, there has been criticism of the patent strategy developed by J. Craig Venter in his project to develop a minimal genome for a living cell, a patent strategy that includes seeking intellectual property in the mechanisms involved in making such synthetic genomes and inserting them into cells. Some suggest that this is an attempt to create a monopoly on what may become basic tools and techniques in the field of synthetic biology. Finally, if narrow patents restrict access to the small building blocks — the parts known as BioBricks™ — and if licences to utilise such parts are costly, this might also inhibit possible future research. There is a fear that if narrow patents in synthetic biology proliferate, they might create what are termed 'patent thickets' or 'anti-commons' — the terms used to describe techniques for blocking future inventions by presenting major financial obstacles in using the materials and information necessary to pursue new lines of innovation (Heller and Eisenberg 1998).

Many synthetic biology groups are considering alternatives to patenting modelled on open source approaches. The most developed case is The Registry of Standard Biological Parts ('The Registry') that is affiliated with iGEM, The BioBricks™ Foundation (BBF) and the recently drafted BioBricks™ Public Agreement (BPA). The Registry is based on the principle of 'get some, give some'. Registry users benefit from using parts and information available in The Registry so they can design and engineer BioBricks™-based biological systems. In exchange, the expectation is that Registry users will contribute back data on existing parts as well as information and new material to BioBricks™, thereby continually improving and growing this community resource.

The Registry's model of openness works well within the context of iGEM. Every year, student teams are given a kit of BioBricks™ from this repository at the beginning of the summer, and are expected to use these parts as they are challenged to design and build a biological system and operate it in living cells; they are also expected to design, test and contribute back new parts to The Registry. The BBF is a non-profit organisation, and it seeks to promote the sharing of biological parts beyond the iGEM, and to develop and implement legal strategies to support and extend this ethos of openness to the synthetic biology community.

As part of the BBF's community mission to spread this openness, leaders behind the BioBricks™ movement have been involved in ongoing discussions of legal options. The publication of the 'BioBricks™ Public Agreement, Draft Version 1a' is the first significant step in this group's effort to establish an appropriately 'balanced' IP system for synthetic biology. According to the authors, the balance is to be achieved by bringing together open access to the parts with the ability to file a patent on inventions created by combinations of those parts to produce a patentable product. The success of this model remains to be seen and its advantages and drawbacks are a matter of current dispute within the synthetic biology community, and among legal and social science scholars.

The International Open Facility Advancing Biotechnology (BIOFAB) is another, more professionalised registry of standard biological parts that is freely available to both academic and commercial users, while also being focused on enabling the rapid design and prototyping of genetic constructs. BIOFAB, however, is still under construction and its intellectual property strategies are likely to evolve over the coming years.[1]

There remain several questions for discussion and debate. For instance, if the synthetic biology community believes it is valuable to maintain the ethos of openness, how can this be fostered in the face of countervailing incentives within some synthetic biology laboratories, which reward individual success and foster hopes of financial rewards for inventions? What modifications might improve the experimental IP formats of current and future repositories of synthetic biology parts, devices, chassis, etc., especially as these grow and develop? How might new repositories of biological parts, developing in different countries and geographical regions, handle IP issues and what kinds of collaboration will they limit or make possible? Protocell construction, for instance, is often left out of synthetic biology's IP discussions — might that sub-field require a whole different set of considerations than either the 'parts, devices, systems' or the 'genome synthesis' approach? How can synthetic biologists, lawyers, social science scholars and industry experts best work together in order to navigate the future evolution of IP in this field?

---

[1] Further examples of 'commons'-based approaches outside of synthetic biology, such as those found in the Biological Innovation for an Open Society (BIOS) initiative, the Tropical Disease Initiative (TDI) and The SNP Consortium (TSC), may be of interest for those wishing to further explore how alternatives of openness have played out in other areas of biotechnology.

Key messages

- Synthetic biology is a 'dual nature' technology, incorporating both software and biotechnology, and hence raising the distinct IP issues that relate to:
  - Software — difficulties in fitting patent and copyright laws; some resolution in open source.
  - Biotechnology — a trend towards increasing patents poses the threat of impeding future innovation because of high costs to access basic materials and information.
- Synthetic biologists are currently experimenting with ways of combining open source approaches with the ability to file patents and direct innovation pathways towards commercialisation. The BioBricks™ Public Agreement is not yet settled and it remains to be seen whether this provides a fruitful form of IP.

Key questions

- How might the field continue to foster a unified ethos when incentives to share may not outweigh countervailing incentives to individual success and financial returns?
- What modifications might improve the experimental IP formats of current and future repositories of synthetic biology parts, devices, chassis, etc., especially as these grow and develop different contents?
- How might IP work in other branches of synthetic biology, such as protocell technology?
- How can synthetic biologists, lawyers, social science scholars and industry experts best work together in order to navigate the future evolution of IP in this field?

9.5 'Playing God' and Challenging the Organism/Machine Divide

Efforts in synthetic biology to construct minimal cells and living organisms raise a number of philosophical and theological questions, including: What is life? Should attempts be made to create new forms of life? Does synthetic biology constitute humans 'playing God'? And if so, is it an inappropriate course of action? Such questions about life and its creation subsequently lead onto many others, for example, how do we distinguish between life and non-life, between the animate and the inanimate, and thus between the organism and the machine? In literature and in history, creatures that have appeared to breach the boundaries between the 'normal' and the 'abnormal' (Canguilhem 2009) or between 'natural' biological organisms and 'designed' technological machines have tended to be seen as monsters, the most famous of which is arguably Frankenstein's monster. Therefore, given that the goal of synthetic biologists is to synthesise 'complex, biologically-based (or inspired) systems, which display functions that do not exist in nature' (NEST 2005) and thus to 'engineer cells into tiny living devices' (Ferber 2004: 158), it is not surprising that synthetic biologists are facing accusations of 'playing God' (ETC 2007) and of treading in the footsteps of Frankenstein (van den Belt 2009). Here we explore both the argument that synthetic biologists are 'playing God' and also the ways in which synthetic biology is challenging the organism/machine divide.

Through the years, scientists such as Copernicus, Newton, Darwin and Einstein have challenged and changed our conceptions of our world and our place within it. Such fundamental challenges continue to confront us as science continues to advance and, with the development of new techniques to 'create' and manipulate life, many of these challenges are now coming from within the life sciences. The potential for humans to act as 'creators' has raised concerns about 'playing God' both in terms of the creation of human life, through processes such as artificial insemination (Kline 1963), and the creation of novel forms of life through genetic engineering (Goodfield 1977). It is into this latter category of creation that synthetic biology falls.

The term 'playing God' is, however, by no means new, having been applied over the years to a wide range of professions and situations. Teachers passing down grades to their students (Skinner 1939), medical teams who make decisions regarding transplant recipients (Harken 1968), the State's application of capital punishment (Gerstein 1960) and even Jane Austen's meddling character Emma (Shannon 1956) have all been charged with, or have faced their own personal concern with, 'playing God'. However, as is also arguably the case with synthetic biology, the charge of 'playing God' in these situations does not necessarily come from those whose opposition is on religious grounds.

Grey (1998), Drees (2002) and Peters (2006) have each highlighted that the charge of 'playing God' is raised by both religious and secular groups. Grey claims, therefore, that rather than always being a religious argument, in a secular context the term is used metaphorically, 'to indicate that the consequences of an act are exceedingly serious or far-reaching and must therefore be considered with very great care,' continuing that, 'the phrase may also be used to describe paternalistic or authoritarian decisions, often resented, made by individuals in positions of power' (Grey 1998: 335). Indeed, even some theologists, such as Peters (2003) and Dabrock (2009), contend that there are no principled objections of a religious nature against the making of new life forms, as synthetic biology aims to do.

Thus, rather than being an explicit religious argument, Davies *et al.* (2009) argue that the term 'playing God' acts as a symbolic expression of inexpressible concerns. Therefore it is potentially more interesting to shift the focus away from the religious wording of the accusation and to look instead at the concerns it represents. 'Playing God' may be better understood, as Grey suggests, as a concern with the use of power, the making of decisions that affect others and with 'humans letting their power and knowledge exceed their caution' (Kirkham 2006: 176). Kirkham claims that the secular version of the term 'playing God' is 'vexing nature' and that the concerns that underlie both are essentially the same. For, as Ball (2010) and Grey (1998) indicate, nature and the natural are often held in high esteem by both religious and secular groups. Thus it is often perceived transgressions against the natural order of things that encounter these objections. As bioethicist Arthur Caplan notes, there is concern that synthetic biologists may be 'manipulating nature without knowing where they are going' (Caplan quoted in Carey 2007: 40).

Caplan also notes that, 'While creating new life may not be playing God, it has revolutionary implications for how we see ourselves. When we can synthesise life, it makes the notion of a living being less special' (Caplan quoted in Carey 2007: 40). Despite the number of textual and pictorial references (see, for example, the cover of *The Economist*, 22–28 May 2010) that would suggest otherwise, Dabrock is not convinced that synthetic biology is indeed about 'creating' new life, either in a religious or a biological sense, yet he agrees with Caplan that synthetic biology is questioning the boundary between the animate and the inanimate (Dabrock 2009).

While the perceived boundary between organisms and machines is a longstanding one, what falls on each side of this divide has differed with time. Descartes, for example, consigned animals to the category of machines, while, before him, Aristotle viewed slaves in the same way (Canguilhem 2009). Arguably we still treat dairy cows as 'milk machines' and many diabetics are now reliant on the insulin produced by *Escherichia coli* or yeast cells — that is to say, living organisms modified to act as insulin-producing 'factories'. Furthermore, humans have a long-standing tradition of challenging this boundary through attempts to create or imitate living organisms (e.g. Leduc 1911; Loeb 1912), and to understand the workings of biological organisms through comparisons with machines and vice versa (Canguilhem 2009).

Nevertheless, the results of such categorisations, imitations and analogies have never truly crossed the divide between the animate and inanimate. Loeb's sea urchins, produced through artificial parthenogenesis were, like babies born following *In Vitro* Fertilisation (IVF), undeniably animate. Leduc's artificial cells, made using inorganic fluids and crystals were, like the products of modern Artificial Intelligence (AI), undoubtedly inanimate. According to Deplazes and Huppenbauer (2009: 56), synthetic biology may well be the first endeavour to successfully blur this boundary, as, 'The aim of producing novel types of living organisms in synthetic biology not only implies the production of living from non-living matter, but also the idea of using living matter and turning it into machines, which are traditionally considered non-living.' Deplazes and Huppenbauer claim that within synthetic biology the boundary between the animate and the inanimate is being approached from both sides, as both natural mechanisms and computers are being used to design the synthetic systems. This suggestion seems to be supported by the recent work of the J. Craig Venter Institute, in which the researchers successfully 'booted up' a synthesised version of the *Mycoplasma mycoides* genome within another cell. At a press conference in May 2010, J. Craig Venter described the resulting organism as 'the first self-replicating species we've had on the planet whose parent is a computer'.

While some hold that there is no significant division between animate and inanimate objects (Haeckel 1866; Zimmer 2008), others see a clear demarcation when it comes to organisms and machines. For example, Canguilhem, a historian and philosopher of science, contended that while the functions of an organism are autonomous, machines are reliant on human intervention. Organisms demonstrate 'self-construction, self-conservation, self-regulation and self-repair', but in order for machines to display these processes they require the 'periodic intervention of human action' (2009: 88). This strict divide is being challenged by synthetic biologists who are attempting to design and build synthetic organisms utilising an engineering approach, but using biological materials. While these synthetic biology 'products' would function biologically, they would be designed and 'built' to perform pre-defined tasks as machines. An example of such a 'biological machine' or 'synthetic organism' is Levskaya *et al.*'s photographic bacteria (2005), which emerged through the iGEM competition — indeed the very name of this competition alone indicates the drive to breach the organism/machine boundary.

As Deplazes and Huppenbauer write, such novel entities 'will affect the concept and evaluation of life and the idea of what constitutes a machine in society and in our culture' (2009: 63). As such, we agree with Dabrock (2009) who contends that synthetic biology should not be given *carte blanche* in its attempts to cross this boundary, and to create new forms of life, and should not regard the concerns expressed under the phrase of 'playing God' as merely theological or rhetorical, but should recognise the central symbolic and cultural issues at stake in its endeavours to tinker with nature, and take a cautious and respectful approach.

Key messages

- Concerns that synthetic biologists are 'playing God' are raised by both secular and religious groups, and such concerns are therefore not necessarily religious in origin.
- The term 'playing God' is symbolic of a wider concern with the appropriate use of power, and in the case of synthetic biology, with manipulating nature.
- Synthetic biology is challenging and blurring the boundary between the animate and the inanimate. This may have consequences for how we define 'life', 'organisms' and 'machines'.

Key questions

- How can we ensure that synthetic biologists use their power to 'tinker' with nature in a gentle and respectful way?
- Does a clear boundary exist between the animate and the inanimate, and if so, does synthetic biology challenge this boundary?
- How do we best describe and understand the 'products' of synthetic biology, as organisms, as machines or as something else?

9.6 Public Value and New Global Inequality

Synthetic biology may contribute to several sectors of the economy, from renewable energy, biosensors, sustainable chemical industries, microbial and plant drug factories to biomedical devices. Two specific lines of development have received considerable attention and investment.

One is the building of a 'sugar economy'. This is to say synthetic biology may drastically transform industries currently relying on fossil fuels (such as paints, cosmetics, plastics and textiles) with affordable and sustainable biofuels made from biological feedstocks (such as agricultural crops, grasses, forest residues, plant oils and algae) (ETC 2008). Bio-Economic Research Associates, a consultancy firm based in Cambridge, Massachusetts, predicts that, with the development of genome synthesis, the biofuels sector could reach $110–150 billion by 2020 (BioERA, 2007). The USA is currently leading the promotion of commercial-scale biofuel productions, and, in January 2010, its Department of Energy pledged a $78 million investment to research algae-based and other advanced biofuels (DOE 2010).

The second key sector is in medical applications. One application that has received a lot of attention is synthetic artemisinin. Artemisinin-based combination therapies (ACT) are recommended by the World Health Organization as the main treatment for uncomplicated malaria (WHO 2010). Traditionally, artemisinin has been extracted from its natural source, the *Artemisia annua* plant, which is endemic to southern China, but also grows in India, Vietnam, Pakistan and South Africa. In July 2010, the San Francisco-based Institute for OneWorld Health (iOWH) announced, however, that research on a semi-synthetic version of artemisinin was to move into full-scale production. Produced by genetically engineered bacteria, this synthetic artemisinin is expected to be cheaper than its plant-derived equivalents.

These examples suggest that one of the key intended public values of synthetic biology is to provide more accessible, sustainable and affordable materials. Yet, as with many other sciences and technologies, the development and applications of synthetic biology in practice may have unintended consequences, including exacerbating global inequalities. Three areas in particular have already attracted social concern: food security, ecological sustainability and economic dislocation.

The production of the first two generations of biofuel, which both rely on growing and harvesting biomass that is then processed, have raised some worries in food security. Some have warned that the financial incentives to grow 'energy crops', such as switchgrass, rather than food, may aggravate food shortages in poor countries (Smil 2003; Doornbosch and Steenblik 2007). Some have suggested that a 'third generation' of biofuels may use techniques from synthetic biology to undertake metabolic engineering on plants, microbes or algae, which would avoid this problem. Such methods might produce plants 'tailored to marginal lands where the soil wouldn't support food crops' (Tollefson 2008: 882), or enable microbes or algae to be grown in vats without diverting land from food crop production (EuropaBio 2008). For example, in the USA, the Joint BioEnergy Institute (JBEI) has identified a three-gene cluster in *Micrococcus luteus* that encodes enzymes that catalyse key steps in the conversion of plant sugars into hydrocarbons. When introduced into *Escherichia coli*, the genes enabled synthesis of long-chain alkene hydrocarbons from glucose. However, problems have also been identified in these third-generation methods: for example, in algal-based methods, critics have pointed to potential contamination and potential need for sugar-rich feedstock, which might reproduce the earlier problems. Thus further attention in addressing the public concerns for the usage of arable land and continuous transparency in the supervision and coordination of economic crop production may continue to be important in the reception and development of a biofuel economy.

Related to concerns about the conversion of land to energy crops are worries about ecological sustainability. Similar to many sciences, such as nanotechnology and genetically modified organisms, the responsible development of a synthetic biology-enabled biofuel economy may require long-term commitment in understanding and communicating its environmental consequences. Appropriate codes of conduct and transnational dialogues amongst regulators, industries and civil society may provide useful tools in addressing concerns such as environmental deterioration and ecological exploitation of poor regions.

Finally, with regard to economic dislocation, there are concerns that the industrial application of synthetic biology may undermine some regions' primary cash crops, and consequently dislocate employees and create pressure for structural change in local economies. One example of this potential is in the production of anti-malaria drugs. The medical effect of artemisinin was first published by Chinese scientists in the 1980s, but as the IP rights for artemisinin are held by Western companies such as Novartis and Sanofi, 'ironically, China has to pay patent fees to foreign countries when Chinese artemisinin products are exported to those countries' (Zhou 2006: 60). In 2009, China's export of artemisinin dropped by 24.2% (*Pharmaceutical Economic News* 2010), and it has been noted that the emergence of synthetic compounds as rival drugs would 'present serious threats and challenges' to the local economy of China's artemisinin-concentrated regions (Su 2010). Meanwhile, Chinese artemisinin companies have indicated that as local industries lack appropriate R&D infrastructure, catching up with Western peers is lengthy and costly. Measurements such as social redistribution to displaced workers, effective economic structural change or adaptation, and necessary financial and regulatory support may therefore be essential in minimising economic impacts on those who are most vulnerable, as the example of artemisinin farmers illustrates. Existing studies on the impact of synthetic products on developing countries have suggested that in cases where synthetic products provide good substitutes for natural ones (as may well be the case with the anti-malaria drug) 'the failure of targeted government intervention can lead to widespread poverty' (Wellhausen and Mukunda 2009: 122).

In summary, while synthetic biology promises a variety of industrial applications, it may also create — although perhaps only temporarily — a polarised situation in which a few leading developing countries are better placed to influence and benefit from this advancement. To minimise new global inequalities, key questions need to be considered — notably which methods should be pursued — and many links in the production chain may require context-specific consideration, such as access to resources (e.g. Who should oversee how land is used?), the use of resources (e.g. How can biomass be acquired responsibly?) and trade adaptation (e.g. How should social welfare be distributed and redistributed?).

Key messages
• Synthetic biology holds great potential in providing society with more accessible, sustainable and affordable materials, but its development and applications may cause unintended new global inequalities, such as food security, ecological sustainability and economic dislocation. • The process of 'industrialising' synthetic biology may therefore require some examination, adaptation and development of existing economic and regulatory structures.
Key questions
• To what extent should those developing applications based on synthetic biology be concerned about the potential consequences in relation to such issues as equity? • Should regulators seek to put measures into effect to protect the most vulnerable from the consequences of development such as synthetic fuels or the replacement of natural products by those produced by corporations using synthetic biology? • Can (and should) those involved in novel research take account of their medium- and long-term social and economic implications?

9.7 Conclusions

There are many ways in which synthetic biology might impact on society, and we have discussed some of these in this chapter. But there are also many ways in which society will impact on synthetic biology. Perhaps the clearest example here is in relation to the issue of regulation.

Since its very early stages, synthetic biologists have played a central role in discussions on how best to regulate the field. This is crucial, as they are the experts on what the science is currently capable of, and of what it might be capable of in the future. Yet, regulation is also about broader, more normative, issues than merely the technicalities of the science. It can be about the level of risk we are willing to accept as a community, and who should bear that burden. Or it can be about how best to share the benefits of the science in a fair and responsible fashion; about the long-term social and environmental impact of the technology and the advisability of promoting its rapid expansion; about respect for nature and where we draw the line for human intervention; about how common resources should be shared etc. Over the last half century, these questions about the regulation of the direction of scientific innovation have been discussed in many ways. Most agree that it is misleading to believe that the direction taken by such innovation is merely an outcome of good science. In our age of 'big science', the direction taken by research and innovation depends on many factors, including the priorities of public funding agencies, or private foundations, of venture capitalists and much more: the decisions taken by each of these at each stage in the process are shaped by social beliefs and expectations as much as by assessments of what is 'good science'. Indeed, such decisions are increasingly shaped by assessments of future economic pay-back from investment, and hence by economic forecasts, horizon scanning and foresight. Given these social factors shaping the direction of science, it is appropriate that these issues are subject to wide public debate, involving synthetic biologists themselves, but also having input from a wider group of stakeholders.

In synthetic biology, as in a number of other 'leading edge' developments such as nanotechnology, public dialogue has become an obligatory part of the process of public funding of scientific and technological innovation. Much of this dialogue has involved soliciting people's views through surveys, focus groups and interviews. However, over the last few years there have been concerted efforts to move towards a more participatory democracy and a more active notion of citizenship when it comes to technology assessment. One example of this is the citizen-based, deliberative 'consensus conference' or 'citizen panel'. These usually involve the recruitment of 10–20 volunteers representing various stakeholders such as interest groups, government, social scientists, practitioners, industry, theologians, clinicians, patients, consumers, security experts, etc. This panel has a preparatory weekend where it is briefed on the issues and where it identifies questions it wants to ask experts. This is then followed by a three- or four-day conference where various experts are invited to answer questions set by the panel, before the panel makes a decision on how it thinks the technology should best be regulated. The idea is that their decision should then be fed into the decision-making process of policymakers.

The extent to which such approaches are effective, or even representative, is often disputed. Thus the debate continues about how to ensure that the concerns of stakeholders are appropriately represented in the decision-making process about the regulation of emerging technologies such as synthetic biology. What is certain, however, is that the science does not develop in a social vacuum; synthetic biology is tightly coupled with society and each affects the other's trajectory. It is vital that the new generation of synthetic biologists, excited about the positive potential of their science and research for the public good, should both recognise, and play a full part in, these deliberative processes.

Reading

Ball, P. (2010). Making life: A comment on 'Playing God in Frankenstein's footsteps: Synthetic biology and the meaning of life' by Henk van den Belt (2009). *Nanoethics* 4: 129–132.

Beatty, D. (1950). 'Sometimes we fail: Whenever we fail to be of help we must look into ourselves. *Pastoral Psychol* 1(7): 16–19.

BioBricks Foundation. *Our Goals*. Available at: http://biobricks.org/Our_Goals.php [Accessed 6 September 2011].

Bio-Economic Research Associates (Bio-ERA). (2007). *Genome Synthesis and Design Futures: Implications for the US Economy*. Bio-ERA, Cambridge, MA.

BIOFAB. *About the biofab*. Available at: http://www.biofab.org/about [Accessed 6 September 2011].

Canguilhem, G. (2009). *Knowledge of Life*. Fordham University Press, New York.

Carey, J. (2007). On the brink of artificial life. *Business Week*. 25 June: 40.
Carlson, R. (2009). The changing economics of DNA synthesis. *Nat Biotechnol* 27: 1091–1094.
Cello, J, Paul, AV and Wimmer, E. (2002). Chemical synthesis of poliovirus cDNA: Generation of infectious virus in the absence of natural template. *Science* 297: 1016–1018.
Central Intelligence Agency. (2003). *The Darker Bioweapons Future*. Unclassified report, released 3 November 2003.
Church, G. (2005). Let us go forth and safely multiply. *Nature* 438: 423.
Dabrock, P. (2009). Playing God? Synthetic biology as a theological and ethical challenge. *Syst Synth Biol* 3: 47–54.
Davies, S, Kearnes, M and Macnaghten, P. (2009). All things weird and scary: Nanotechnology, theology and cultural resources. *Culture and Religion* 10(2): 201–220.
Department of Energy, USA. (2010). Secretary Chu announces nearly $80 million investment for advanced biofuels research and fueling infrastructure. Press release, 13 January 2010.
Deplazes, A and Huppenbauer, M. (2009). Synthetic organisms and living machines: Positioning the products of synthetic biology at the borderline between living and non-living matter. *Syst Synth Biol* 3: 55–63.
Doornbosch, R and Steenblik, R. (2007). Biofuels: Is the cure worse than the disease? *OECD Round Table on Sustainable Development*. 11–12 September 2007, OECD, Paris.
Drees, W. (2002). 'Playing God? Yes!' Religion in the light of technology. *Zygon: Journal of Religion and Science* 37: 643–654.
Editorial (2010). Artemisinin export slumped, the industry in danger. *Pharmaceutical Economic News* (China). 20 January.
ETC Group. (2007). *Extreme Genetic Engineering — An Introduction to Synthetic Biology*, January. Available at: http://www.etcgroup.org/en/node/602 [Accessed 6 September 2011].
ETC Group. (2008). *Commodifying Nature's Last Straw? Extreme Genetic Engineering and the Post-Petroleum Sugar Economy*, October. Available at: http://www.etcgroup.org/en/node/703 [Accessed 6 September 2011].
EuropaBio. (2008). *EuropaBio Fact Sheet: Biofuels and Land Use*, March. EuropaBio, Brussels.
Ferber, D. (2004). Synthetic biology. Time for a synthetic biology Asilomar? *Science* 303: 159.
Fletcher, J. (1960). The patient's right to die. *Harper's Magazine*. October: 139–143.
Gerstein, R. (1960). A prosecutor looks at capital punishment. *J Crim Law Crimnol* 51: 252–256.
Goodfield, J. (1977). *Playing God: Genetic Engineering and the Manipulation of Life*. Random House, New York.
Grey, W. (1998). 'Playing God' in Chadwick, R. (Ed.). *The Concise Encyclopaedia of the Ethics of New Technologies*. Academic Press, San Diego, 335–339.
Haeckel, E. (1866). *Generelle Morphologie der Organismen* (*General Morphology of the Organisms*). Reimer, Berlin.
Harken, D. (1968). One surgeon looks at human heart transplantation. *Dis Chest* 54(4): 349–352.
Heller, M. (1998). The tragedy of the anticommons: Property in the transition from Marx to markets. *Harvard Law Rev* 111: 621–688.
Heller, M and Eisenberg, R. (1998).Can patents deter innovation? The anticommons in biomedical research. *Science* 289: 698–701.
Hope, J. (2008). *Biobazaar: The Open Source Revolution and Biotechnology*. Harvard University Press, Cambridge, MA.
iGEM. (2010). *About*. Available at: http://2010.igem.org/About [Accessed 6 September 2011].
Institute for OneWorld Health. (2010). The drug, produced by genetically engineered bacteria, is much cheaper than the plant-derived drug available today. Press release, 7 July 2010.
Kirkham, G. (2006). 'Playing God' and 'vexing nature': A cultural perspective. *Environ Value* 15: 173–195.
Kline, C. (1963). The potential significance of sperm banks. *Advances in Sex Research* 1: 123–127.

Kumar, S and Rai, A. (2007). Synthetic biology: The intellectual property puzzle. *Tex Law Rev* 85: 1744–1768.

Leduc, S. (1911). *The Mechanism of Life*. William Heinemann, London.

Lentzos, F, Bennett, G, Boeke, J, *et al.* (2008). Visions and challenges in redesigning life. *BioSocieties* 3(3): 311–323.

Levskaya, A, Chevalier, A, Tabor, J, *et al.* (2005). Engineering *Escherichia coli* to see light. *Nature* 438: 441–442.

Loeb, J. (1912). *The Mechanistic Conception of Life*. University of Chicago Press, Chicago.

McGregor, D. (1957). An uneasy look at performance appraisal. *Harvard Bus Rev* 35: 89–94.

National Research Council. (2004). *Biotechnology in an Age of Terrorism*. National Academies Press, Washington DC.

National Research Council. (2009). *A Survey of Attitudes and Actions on Dual Use Research in the Life Sciences*. National Academies Press, Washington DC.

New and Emerging Science and Technology. (2005). *Synthetic Biology: Applying Engineering to Biology. Report of a NEST High-Level Expert Group*. European Commission, Luxembourg.

OpenWetWare. A wiki-style online community for bioscience. Available at: http://openwetware.org/wiki [Accessed 6 September 2011].

Peters, T. (2003). *Playing God? Genetic Determinism and Human Freedom*. Routledge, New York.

Peters, T. (2006). 'Contributions from practical theology and ethics' in Clayton, P and Simpson, Z. (Eds). *The Oxford Handbook of Religion and Science*. Oxford University Press, Oxford, 372–387.

Rai, A and Boyle, J. (2007). Synthetic biology: Caught between property rights, the public domain, and the commons. *PLoS Biology* 5: 389–393.

Registry of Standard Biological Parts. Available at: http://partsregistry.org [Accessed 6 September 2011].

Schmidt, M, Torgersen, H, Ganguli-Mitra, A, *et al.*(2008). SYNBIOSAFE e-conference: Online community discussion on the societal aspects of synthetic biology. *Syst Synth Biol* 2: 7–17.

Shannon, E. (1956). Emma: Character and construction. *PMLA* 71(4): 637–650.

Skinner, M. (1939). Playing God. *Engl J* 28(6): 471.

Smil, V. (2003). *Energy at the Crossroads*. MIT Press, Cambridge, MA.

Smith, GL and Davison, N. (2010). Assessing the spectrum of biological risks. *Bull Atomic Scientists* 66(1) 1–11.

SPB. (2006). *Criteria for Sustainable Biomass Production*. 14 July 2006. Project Group for Sustainable Production of Biomass.

Su, K. (2010). Blind expansion on Artemisinin. *Pharmaceutical Economics News* (China). 30 March.

Tollefson, J. (2008). Energy: Not your father's biofuels. *Nature* 451: 880–883.

Tucker, JB and Zilinskas, RA. (2006). The promise and perils of synthetic biology. *The New Atlantis* 12: 25–45.

Tumpey, TM, Basler, CF, Aguilar, PV, *et al.* (2005). Characterization of the reconstructed 1918 Spanish influenza pandemic virus. *Science* 310: 77–80.

van den Belt, H. (2009). Playing God in Frankenstein's footsteps: Synthetic biology and the meaning of life. *Nanoethics* 3: 257–268.

Vogel, KM. (2008). Framing biosecurity: An alternative to the biotech revolution model? *Science and Public Policy* 35: 1.

Wellhausen, R and Mukunda, G. (2009). Aspects of the political economy of development of synthetic biology. *Syst Synth Biol* 3: 115–123.

World Health Organization. (2010). Drug resistance could set back malaria control success. Press release, 25 February 2010.

World Intellectual Property Organization: Available at: http://www.wipo.int/portal/index.html.en [Accessed 6 September 2011].

World Intellectual Property Organization (undated). *Frequently Asked Questions*. Available at: http://www.wipo.int/patentscope/en/patents_faq.html#patent [Accessed 6 September 2011].

World Intellectual Property Organization (undated). *Understanding Copyright and Related Rights*. Available at: http://www.wipo.int/freepublications/en/intproperty/909/wipo_pub_909.html [Accessed 6 September 2011].

Zhou, EY. (2006). Chinese biogenerics and protection of IP: Large market potential and solid foundation set the stage. *Genetic Eng Biotechn* 26(15): 56–60.

Zimmer, C. (2008). Artificial life? Old news. Wired.com. Available at: http://www.wired.com/science/discoveries/commentary/dissection/2008/01/dissection_0125 [Accessed 25 January 2008].

APPENDIX 1

Proforma of Common Laboratory Techniques

Technique: Polymerase Chain Reaction (PCR)

Inventor: Kary Mullis (1983)

Molecule: DNA

Basic Concept: PCR is a technique that is used to make many copies of DNA for cloning and other purposes. Short single-stranded pieces of DNA called primers are designed in regions flanking the sequence to be copied (also called the DNA template). These are then synthesised, usually by a commercial company. They are added to the reaction mixture along with free nucleotide triphosphates, a temperature stable DNA polymerase, the DNA template and buffering components. PCR follows a series of steps at different temperatures which are repeated over and over again to give a high yield of the DNA of interest. In the first step, the reaction mixture is heated to a high temperature which denatures the template forming single strands. The temperature is then lowered and the primers bind (anneal) to the designed region, one to each strand of DNA. The temperature is then increased to the optimum working temperature of the DNA polymerase and new DNA is synthesised. This process is then repeated to make further copies. Because each copy of the template results in two synthesised copies per cycle, the amplification of the DNA proceeds exponentially.

Uses in Synthetic Biology: The most common use of PCR in synthetic biology is to create many copies of DNA parts for cloning and/or assembling devices and systems. It is also used in some cases for quantitation (see quantitative PCR below).

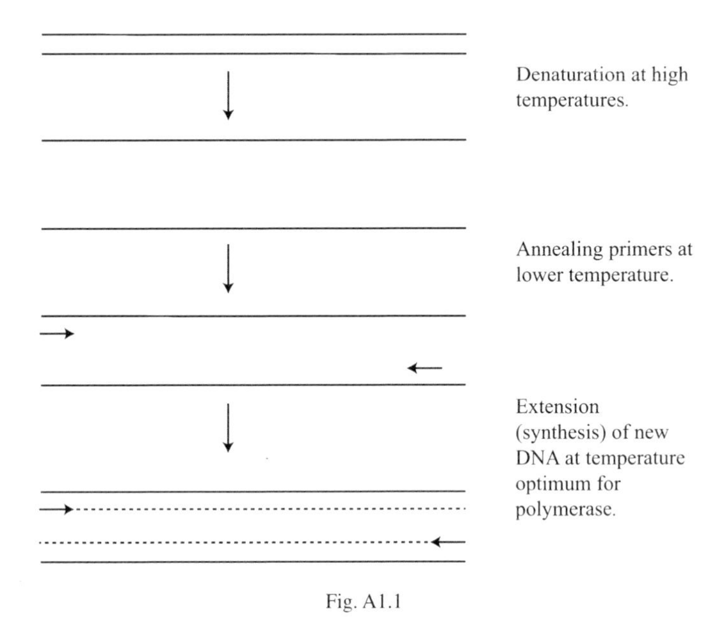
Denaturation at high temperatures.
Annealing primers at lower temperature.
Extension (synthesis) of new DNA at temperature optimum for polymerase.

Fig. A1.1

Technique: Restriction Enzyme Cloning

Inventor: Herbert Boyer and Stanley Cohen (1973)

Molecule: DNA

Basic Concept: Cloning is a series of molecular biology steps whereby DNA of interest is inserted into a host cell or chassis. It begins with PCR amplification of the DNA. The DNA is then treated with enzymes which recognise and cut specific sequences, leaving behind a short single-stranded DNA overhang. A vector, or circular piece of carrier DNA, is likewise treated to create complementary ends. These are then annealed together and joined by an enzyme called ligase to recreate a fully circularised piece of DNA (a vector now containing the piece of DNA of interest as an insert). Cells are then induced to take up the circularised plasmids in a process called transformation (for prokaryotes and yeast) or transfection (for mammalian cells). Including an antibiotic resistance marker or another gene which gives cells an advantage allows the user to selectively keep the cells which have taken the plasmid, creating a population of clones.

Uses in Synthetic Biology: Restriction enzyme cloning forms the basis of the BioBrick™ assembly method, which follows a similar flowsheet but uses specific restriction enzymes. It is also the workhorse of traditional molecular biology. However, restriction enzyme cloning is slow and tedious because it is composed of multiple steps which must be carried out for each piece of DNA that is added to a system. In the future, it is envisioned that new DNA assembly techniques will supplant the multistep cloning techniques enabling faster synthesis of complex systems.

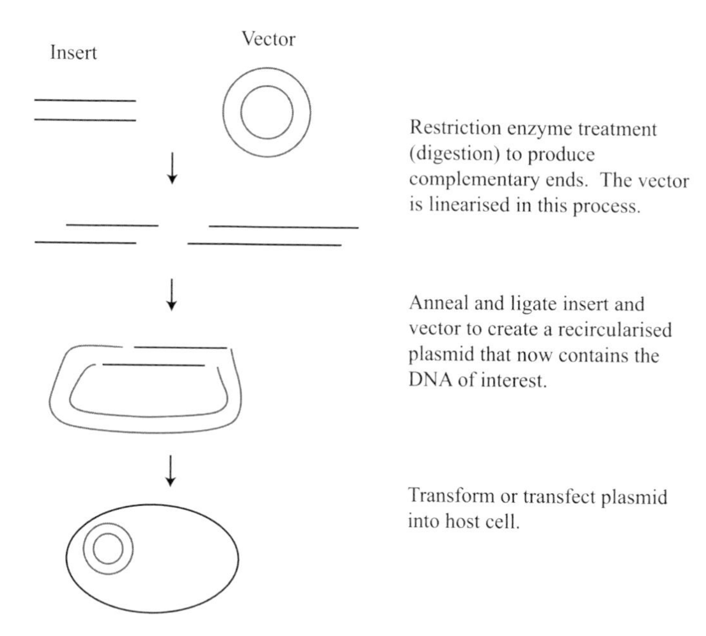

Insert
Vector
Restriction enzyme treatment (digestion) to produce complementary ends. The vector is linearised in this process.
Anneal and ligate insert and vector to create a recircularised plasmid that now contains the DNA of interest.
Transform or transfect plasmid into host cell.

Fig. A1.2

Technique: BioBrick™ Assembly

Inventor: Tom Knight (2003)

Molecule: DNA

Basic Concept: BioBrick™ assembly is a version of restriction enzyme cloning where specific restriction enzymes are used in the process in order to create re-usable restriction sites that allow for sequential assembly of a large number of parts. Each BioBrick™ part is flanked by two different restriction enzyme sites on the 5' and 3' ends of the DNA. Upstream of the part are the *Eco*RI and *Xba*I sites and downstream are the *Pst*I and *Spe*I sites. The destination vector also contains all four of the sites. When an empty vector is used as the destination, then, in principle, any of the restriction sites can be used for cloning the first part, but the power of the BioBrick™ method is that vectors that already contain parts can still be used to add additional pieces to the device. To clone a new part upstream of an existing assembly, the part is digested with *Eco*RI and *Spe*I restriction enzymes and the vector is digested with *Eco*RI and *Xba*I. Since *Xba*I and *Spe*I produce compatible sticky ends, the part can be ligated into the vector by annealing the two *Eco*RI overhangs of the part and vector and the *Xba*I overhang of the vector with the *Spe*I overhang of the insert. Once ligated the combined *Xba*I–*Spe*I site cannot be redigested with either restriction enzyme, so new parts can be added sequentially in the same manner. Similarly, to ligate a part into the vector downstream of an existing assembly, the part is digested with *Xba*I and *Pst*I and the vector with *Pst*I and *Spe*I.

Uses in Synthetic Biology: BioBrick™ assembly is the basis of construction of many of the devices and systems in the iGEM competition and it is also fairly heavily used in synthetic biology projects in general. However, because it relies on restriction enzymes and has the same associated disadvantages, it is likely that BioBrick™ cloning will be replaced by other DNA assembly strategies in the future.

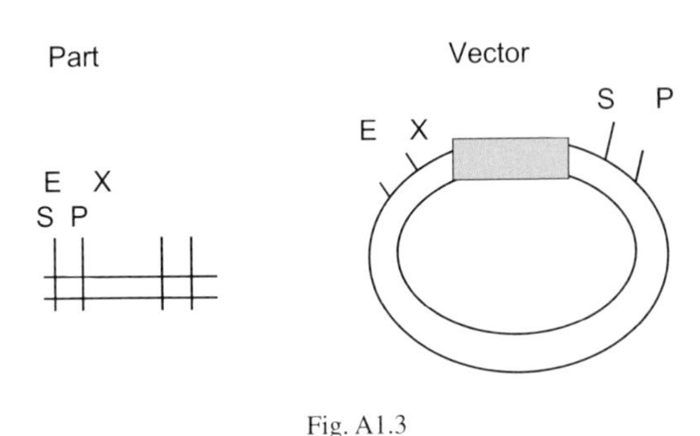

Fig. A1.3

Technique: Gel electrophoresis

Molecule: DNA, RNA or Protein

Basic Concept: Gel electrophoresis is a sample analysis technique in which a sample is added to a matrix and the components are separated on the basis of charge/mass ratio using an electric field. The speed of the migration depends on the size of the particle as well as the concentration of the matrix that composes the gel. A standard containing molecules of known size and concentration is usually added to the gel in order to aid in interpretation of the results. DNA gel electrophoresis most commonly uses agarose as a matrix, although other matrices can be used, particularly for small or single-stranded products. Since DNA is highly negatively charged due to the phosphate backbone, it will migrate towards positive electrodes when voltage is applied and the speed of migration is directly proportional to its size. RNA gel electrophoresis is very similar to DNA gel electrophoresis and also contains a phosphate backbone which provides a negative charge. Protein gel electrophoresis uses polyacrylamide as a matrix. However, in contrast to nucleic acids, proteins do not have a uniform charge. Therefore, they will migrate based on a combination of charge and mass unless reacted with chemicals that alter their charge. For routine analyses to determine size and/or concentration of proteins, samples are usually denatured and reacted with sodium dodecyl sulphate (SDS), a detergent which coats the protein with negative charge so that it separates according to mass just like on a nucleic acid gel. Proteins can also be electrophoresed on polyacrylamide gels without SDS treatment; these are known as native gels.

Uses in Synthetic Biology: Gels are a very common method of sample analysis. DNA agarose gels are used to check the progress during cloning experiments, often after every step and for rough quantification of the DNA concentration of samples during assembly of systems. RNA gels are used to check the purity and concentration of samples before further steps such as quantitative PCR (see below). Protein gels can be used to verify that proteins are expressed and/or to quantify when paired with Western blot (see below).

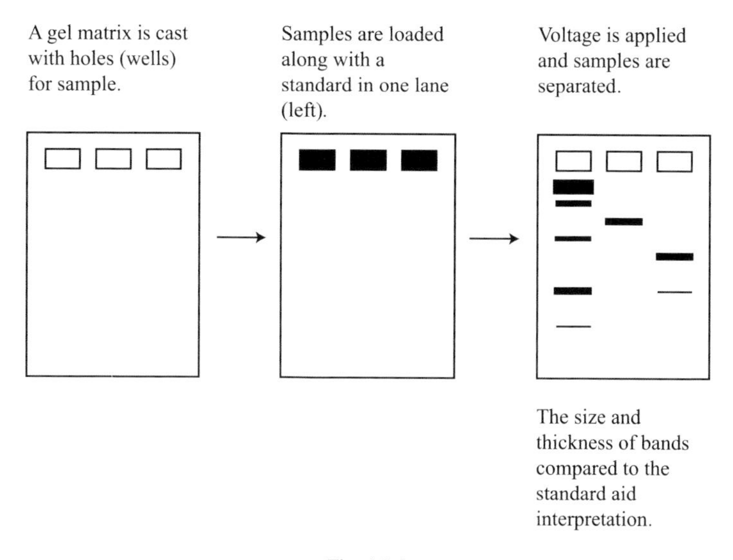

A gel matrix is cast with holes (wells) for sample.
Samples are loaded along with a standard in one lane (left).
Voltage is applied and samples are separated.
The size and thickness of bands compared to the standard aid interpretation.

Fig. A1.4

Technique: Sequencing

Inventor: Fredrick Sanger (1974)

Molecule: DNA

Basic Concept: Current DNA sequencing is based on a method developed by Fredrick Sanger. At its heart is a PCR amplification reaction similar to what is described above, but with modified nucleotides mixed in. These nucleotides lack the necessary chemical group to enable further chain extension and thus cause early termination of the amplification. They are also labelled with fluorescent dyes which allow the sequencer to know the identity of the base which is last in the sequence. The PCR reaction results in a series of shorter DNA fragments, each labelled at the final base. These can then be sorted by size and the sequence 'read' from the identity of the last base via the fluorescent label. In most laboratories, DNA sequencing is sent out to a professional laboratory which performs the sequencing and returns a result by email.

Uses in Synthetic Biology: Sequencing is used to confirm the identity of cloned DNA as well as to ensure that no errors were introduced in the cloning or assembly process.

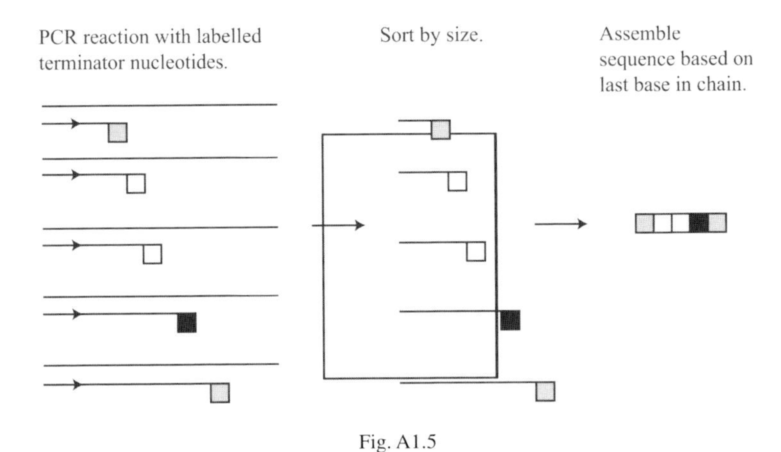

PCR reaction with labelled terminator nucleotides.
Sort by size.
Assemble sequence based on last base in chain.

Fig. A1.5

Technique: Southern blot

Inventor: Edwin Southern (1975)

Molecule: DNA

Basic Concept: Southern blotting is used to transfer DNA that has been electrophoresed to a membrane where it can be hybridised with fluorescent probes. The probes are designed to bind to specific DNA sequences, making the technique useful for detecting when a specific sequence is present in a sample. The probe signal can also be used to roughly quantify the amount of DNA present.

Uses in Synthetic Biology: Southern blotting is one method to determine copy number of a gene, but is very rarely used in synthetic biology. It has been largely replaced by quantitative PCR (see below).

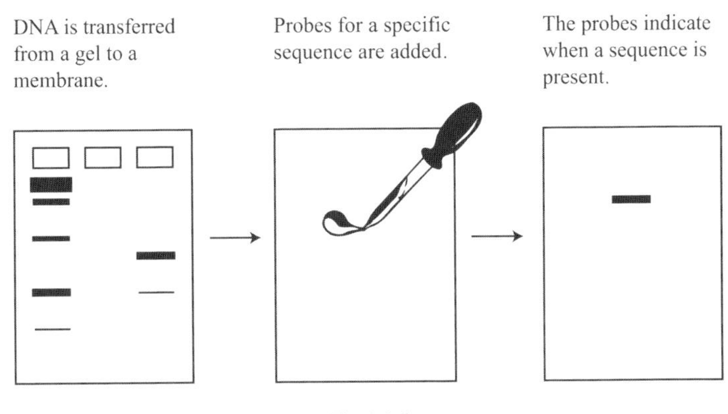

DNA is transferred from a gel to a membrane.
Probes for a specific sequence are added.
The probes indicate when a sequence is present.

Fig. A1.6

Technique: Quantitative PCR (Real-Time PCR)

Molecule: DNA

Basic Concept: Quantitative PCR is very similar to the standard PCR reaction described above. In addition to the normal reagents the reaction mixture also contains a fluorescent dye which intercalates into the DNA that is formed. This allows a measurement of DNA concentration (in comparison to a standard curve of known DNA concentration).

Uses in Synthetic Biology: This is the preferred method of measuring DNA copy number. It can also be combined with reverse transcription (see below) to measure mRNA copy number and to track gene expression changes in response to a genetic circuit.

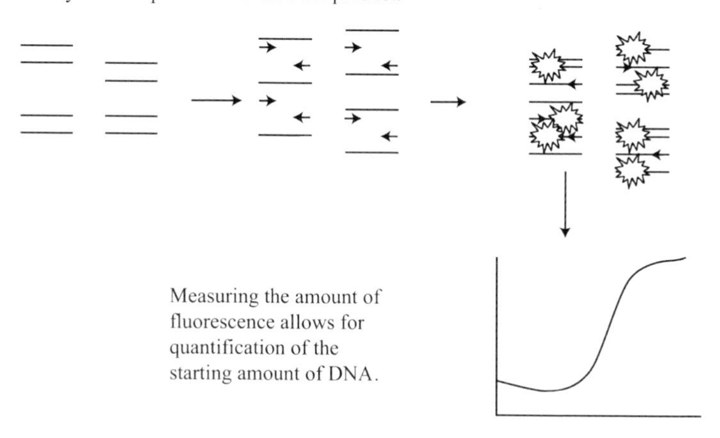

DNA template is subjected to PCR in the presence of fluorescent dye.
The dye is incorporated into the PCR product.
Measuring the amount of fluorescence allows for quantification of the starting amount of DNA.

Fig. A1.7

Technique: Northern blot

Inventor: James Alwine, David Kemp and George Stark (1977)

Molecule: RNA

Basic Concept: Northern blotting is the same procedure as Southern blotting, applied to RNA molecules. RNA is separated by gel electrophoresis, transferred to a membrane and then hybridised with probes designed to detect the desired sequence. Its name is a take on Southern blotting.

Uses in Synthetic Biology: Northern blotting can be used to detect and quantify RNA, but since it is a difficult and time-consuming procedure it has largely been replaced by quantitative PCR.

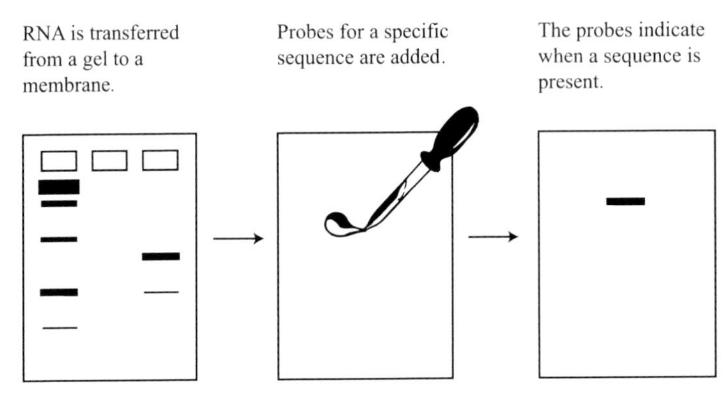
RNA is transferred from a gel to a membrane.
Probes for a specific sequence are added.
The probes indicate when a sequence is present.

Fig. A1.8

Technique: Microarray

Molecule: DNA (though versions for other molecules exist)

Basic Concept: A microarray is very similar to a Southern blot run in reverse. A large number of probes are printed on a chip surface and a DNA sample is applied to the chip. The DNA binds to probes of complementary sequence, giving a signal.

Uses in Synthetic Biology: Microarrays can be used to assay expression changes of a large number of genes. RNA is extracted and then converted to single-stranded DNA (cDNA) using a reverse transcription reaction (see RT-PCR below) which is applied to the chip. This is useful for detecting systems level changes in gene regulation response to a synthetic circuit.

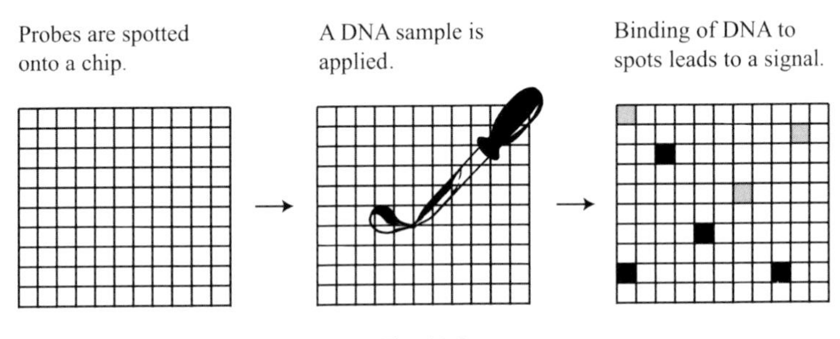
Probes are spotted onto a chip.
A DNA sample is applied.
Binding of DNA to spots leads to a signal.

Fig. A1.9

Technique: Reverse transcription

Molecule: RNA

Basic Concept: Reverse transcription is used to convert RNA to single-stranded DNA, also called complement DNA or cDNA. This enables long-term storage as well as the application of any methods that utilise DNA (e.g. microarray, quantitative PCR) to an RNA sample. A viral enzyme, reverse transcriptase, is used to copy the mRNA to a single strand of DNA.

Uses in Synthetic Biology: Reverse transcription coupled with quantitative PCR is the preferred method for quantifying mRNA copy number in synthetic systems.

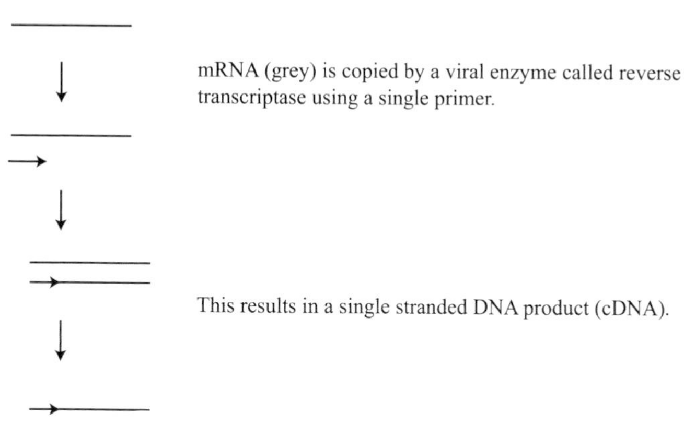

mRNA (grey) is copied by a viral enzyme called reverse transcriptase using a single primer.
This results in a single stranded DNA product (cDNA).

Fig. A1.10

Technique: Western blot

Inventor: George Stark and J Gordon (1979), W Neal Burnette (1981)

Molecule: Protein

Basic Concept: Western blotting is a procedure similar to Southern blotting, but applied to proteins as the molecule of interest. Western blotting uses antibodies as the molecule of detection rather than labelled nucleotides. After separating the sample by gel electrophoresis, the protein bands are transferred to a membrane which is incubated with an antibody that binds the protein of interest as its antigen. This antibody can be labelled (direct detection) or else a secondary antibody can be used to detect where the first antibody has bound. The secondary antibody will be against generic sequences from the organism in which the primary antibody is made (e.g. anti-mouse antibodies).

Uses in Synthetic Biology: Western blotting is used to determine whether an organism is expressing a protein of interest and can be used to quantify the amount of that protein expressed. In contrast to the two other types of blotting, it is still heavily used.

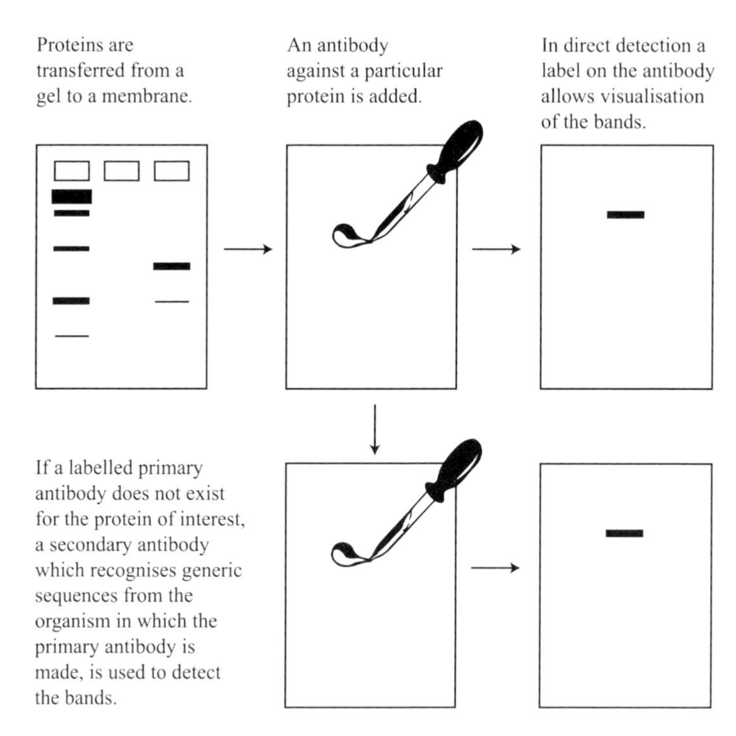
Proteins are transferred from a gel to a membrane.
An antibody against a particular protein is added.
In direct detection a label on the antibody allows visualisation of the bands.
If a labelled primary antibody does not exist for the protein of interest, a secondary antibody which recognises generic sequences from the organism in which the primary antibody is made, is used to detect the bands.

Fig. A1.11

APPENDIX 2

Glossary

Abstraction: A process where higher concepts are built from combining concepts that themselves have been founded on first principles. The purpose of abstraction is to hide unnecessary details and manage complexity by allowing individuals to work at any one level of complexity without regard for the details that define other levels, while simultaneously allowing for the principled exchange of limited information across levels.

Activator: A protein which binds to DNA in the region of the promoter (usually upstream) and aids in establishing transcription.

Akaike Information Criterion (AIC): A criterion for model selection among a class of parametric models with different numbers of parameters. Loosely speaking the AIC provides a quantitative measure of the compromise between accuracy and complexity of the model.

Amino acid: Building block of proteins.

Amplitude: The height of the peak obtained in an oscillation.

AND gate: A logic gate where all inputs must be present to produce the output.

Anticodon: The portion of the tRNA that base pairs with the mRNA codon to position the correct amino acid for protein synthesis.

Antisense (strand): The complementary strand to the sense strand. It does not actually encode for a gene, but its sequence is determined through base pairing.

Base pair: A single nucleotide in a macromolecule. The same term is sometimes used for the actual physical hydrogen bonding between two nucleic acid strands.

Bayesian Information Criterion (BIC): A criterion for model selection among a class of parametric models with different numbers of parameters. It is a variant of the Akaike Information Criterion, whose development is based on a Bayesian argument.

Biobricks™: DNA biological parts with standardised prefix and suffix DNA sequences that allow them to be routinely assembled.

Bioerror: Accidental release of biological organisms.

Biohacking: Synthetic biology experiments by non-experts, analogous to computer hacking.

Biosafety: Measures to protect against accidental exposure to biological organisms.

Biosecurity: Measures to counteract bioterrorism.

Bioterrorism: The deliberate release of biological organisms with the intent to cause harm.

Bottom-up: Method of engineering in which a system is built up from scratch beginning with basic elements or building blocks.

Capillary electrophoresis: A fast, miniaturised method for gel electrophoresis where hundreds of thin capillary tubes contain gel, replacing the need for large slab gels.

Central dogma: The central tenet of information flow in molecular biology that DNA is transcribed to RNA which is translated into proteins.

Chassis: A chassis is the underlying biology in which a device or system is implemented. This can be a living organism (host) or an *in vitro* system for transcription and translation.

Chemical master equation: Differential equation describing the time evolution of the probability of a system to occupy each one of a discrete set of possible chemical states.

Chimera: A fusion of genes from two different organisms.

Coarse-grained models: A description of a system which regards large subcomponents without giving details about their internal description. On the contrary, **fine-grained models** additionally provide details of the smaller components of which the larger ones are composed.

Coding Region: See **open reading frame**.

Codon optimisation: For most amino acids there are multiple codons. Different codons for the same amino acid can have a variety of efficiencies during translation of an mRNA into a protein, and these efficiencies vary across nature. Efficiency of translation can be improved by selecting the optimal codons for the chassis cell.

Codons: In a protein-coding sequence each amino acid of the protein is encoded by three bases of mRNA, which are transcribed from the equivalent three bases of DNA in the gene.

Colourimetry: Quantitative measurement of a colour change.

Commodity (chemicals): Molecules that are made in very large scale and usually sold for a very small profit per unit, but are sold in such volume that the process is still economical.

Complement/Complementarity: The 'opposite' strand of nucleic acid whose sequence is determined by base pairing with the sense strand.

Composable design models: Models useful for design purposes that can be easily arranged or composed together to form super-models of higher complexity.

Computer-Aided Design (CAD): The use of computer technology for the process of design and design-documentation. CAD software or environments provide the user with input-tools for the purpose of streamlining design processes.

Consensus: The most common nucleic acid base or amino acid in a particular position when a group of sequences are compared.

Conserved (sequence): Areas of nucleic acid or protein that do not vary between organisms.

Constitutive (promoter): A promoter from which transcription is constantly taking place.

Context dependence: The idea that the nucleic acid sequence in the immediate vicinity of the parts can affect its function.

Degenerate (code): The idea that since there are more triplet codons than amino acids required, some amino acids will be encoded by more than one codon.

Deterministic model: Mathematical model in which outcomes are precisely determined through known relationships among states and events, without any room for random variation. In a deterministic model, every set of variable states is uniquely determined by parameters in the model and by sets of previous states of these variables. Therefore, deterministic models perform the same way for a given set of initial conditions.

Device: A device is a collection of parts that perform a higher order function, usually human-defined.

DNA: Deoxyribonucleic acid. A biological macromolecule that encodes the information necessary for an organism to function. Consists of a deoxyribose (sugar)-phosphate backbone and four bases: Adenine, Thymine, Cytosine and Guanine.

Downstream (DNA): Part of the DNA that is closer to the 3' end of the strand than the reference sequence.

Dual use: The idea that scientific discoveries can also be used for other, unintended, purposes such as military applications by rogue states.

Eukaryote: Cells that contain a true nucleus and, often, other compartments (organelles) with specific functions.

Extrinsic noise: Noise due to variability in the amount or state of biochemical species external to the considered biochemical process, e.g. transcription or translation.

Feedback: A phenomenon whereby the output of a circuit influences the future performance of that circuit. Feedback can either be positive or negative.

Fine chemicals: Speciality chemicals, often with high retail value, which are not made on a very large scale. For example, most pharmaceutical drugs are fine chemicals.

Flow cytometry: Fast measurement of colour, fluorescence and luminescence from thousands of single cells, achieved by passing cells in a thin capillary tube at high speed past sensitive detectors.

Fluorometry: Quantitative measurement of light emitted in a fluorescence reaction.

Fluorophores: A molecule that fluoresces by emitting light after being stimulated by a different wavelength of light.

Frequency: See **periodicity**.

Functional genomics: The use of genome sequence data to describe molecular biology functions in cells, such as transcription regulation.

Gene synthesis: The construction of double-stranded DNA over 200 bp long from chemicals.

Genome Engineering: The rational re-writing, editing or complete novel design of whole genomes.

Granularity: The scale or level of detail present in a mathematical model (see also **coarse-grained**, above).

Graphical User Interface (GUI): User interface that allows users to interact with programs in more ways than typing. The actions are usually performed through direct manipulation of the graphical elements such as clicking on an icon.

Heterologous: Non-native, coming from a different organism.

High-throughput techniques: Experimental methods that process hundreds or thousands of samples in parallel, rather than one at a time.

High-value added: Molecules which can be sold for much more than the cost of the starting materials.

Holoenzyme: A complete enzyme that contains all of its constituent parts and is, therefore, functional.

Human practices: A field which considers the societal aspects of technology including the impact on the end-user.

Identifiability: Property which a model must satisfy in order for inference to be possible. A model is **identifiable** if it is theoretically possible to learn the true value of this model's underlying parameter after obtaining an infinite number of observations from it. Usually the model is identifiable only under certain technical restrictions, in which case the set of these requirements is called the **identification conditions**.

iGEM: The International Genetically Engineering Machine synthetic biology competition in which teams of students design, model, implement and then present a new synthetic biology system or circuit at an organised event (Jamboree).

***In silico*:** Performed on computer or via computer simulation.

***In vitro*:** Performed not in a living organism but in a controlled environment, such as in a test tube or Petri dish.

***In vivo*:** Performed in live cells or living organisms.

Inducible (Promoter): A common term for a regulated promoter where the repressor is present in sufficient numbers that the default state of the promoter is 'off'. When a small molecule is added to titrate the repressor away from its binding site transcription can then proceed.

Intellectual property: Term for an invention or an idea which is new and therefore qualifies for patent, copyright or trademark protection.

Intergenic: Region between genes.

Intrinsic noise: Noise due to the inherent stochasticity of biochemical processes such as transcription and translation.

Inverter: Another term for **NOT gate** (see below).

Isocaudomers: Pairs of restriction enzymes that have slightly different recognition sequences but the DNA they cut can still link together naturally because of compatible overhanging ends.

Likelihood Ratio Test (LRT): A test to compare the fit of two models, one of which is nested within the other. This often occurs when testing whether a simplifying assumption for a model is valid, as when two or more model parameters are assumed to be related.

Logic gate: A device that relies on multiple inputs to control a single output, the basis of modern computing.

Luminescence: The light-emitting property of a molecule that can convert a chemical reaction into one or more photons.

Luminometry: Quantitative measurement of a light-emitting reaction.

Microbe: A single-celled, microscopic organism.

Microfluidics: The precise control and manipulation of fluids and reactions in tubes and vessels typically micrometres in size.

Microscopy: The analysis and measurement of cells using optical microscopes.

Minimal cells: Cell systems specifically designed and engineered to exist with a minimal amount of genetic information.

Model: A simplified or idealised description or conception of a particular system, situation or process, often in mathematical terms, that is put forward as a basis for theoretical or empirical understanding, or for calculations, predictions, etc.

Modular/modularity: Modularity is a property of a system which allows individual parts or devices to be exchanged without affecting the function of the parts or devices which remain untouched.

mRNA: Messenger RNA is produced from a gene by transcription. It is recognised and translated by a ribosome to create the protein encoded by that gene.

mRNA lifetime: The lifetime of an mRNA is a measure of the amount of time the mRNA exists in a cell after it has been produced. During this time it can repeatedly be translated to make many copies of the protein it encodes.

Mutation and selection: The main method of natural evolution and directed evolution, where mutations occur in DNA sequence and lead to changes in cell behaviour that can be selected for.

NAND Gate: Contraction of **NOT** and **AND**, a NAND gate produces an output signal except for in the presence of two inputs. It shows the opposite behaviour of an AND gate.

NOR Gate: Contraction of **NOT** and **OR**, a NOR gate produces an output signal only when both inputs are absent, i.e. if NOT A NOR B, then 'ON', else 'OFF'.

NOT Gate: Also called an inverter. A NOT gate inverts a signal. In the presence of an input signal it does not produce an output. However, in the absence of a signal it does produce an output.

Open reading frame: See also **protein coding sequence**. A DNA sequence that contains the necessary cues for translation of a protein.

Open source: An idea borrowed from software engineering, where the 'code' should be shared among users in order to promote innovation.

Operator: A DNA sequence that is recognised by a repressor or activator of transcription.

Operon: A set of open reading frames, often part of a single metabolic pathway, transcribed from a single promoter.

OR gate: A logic gate which produces an output signal if one of the input signals is present, i.e. if A OR B, then 'ON', else 'OFF'.

Ordinary differential equation: A relation that contains functions of only one independent variable (typically time), and one or more of their derivatives with respect to that variable.

Orthogonal/orthogonality: When two or more similar systems are engineered so that they cannot interact with each other.

Palindrome (Palindromic): A sequence that is the same when read from left to right on the sense strand or right to left on the antisense strand (for example $\begin{smallmatrix}\text{AAAATTTT}\\ \text{TTTTAAAA}\end{smallmatrix}$). This has the effect that complementary base pairs within a single strand can hydrogen bond, forming a hairpin structure.

Parsimony or Occam's Razor Principle: Principle which generally recommends selecting the competing hypothesis that makes the fewest new assumptions when the hypotheses are equal in other respects. The Razor is a principle that suggests we should tend towards simpler theories until we can trade some simplicity for increased explanatory power/accurateness.

Part: A piece of DNA which encodes a biological function. Examples include a promoter, a ribosome binding site or a gene-coding region.

Partial differential equation: A relation involving an unknown function (or functions) of several independent variables (typically time and space) and their partial derivatives with respect to those variables.

Periodicity: The length of time elapsed from peak to peak in an oscillatory system. Also called the **frequency**.

Platform chemicals: Chemical building blocks that serve as the starting point for a variety of useful molecules.

Polymerase chain reaction (PCR) methods: Molecular biology techniques that use the polymerase chain reaction (PCR) to amplify, modify and link together DNA sequences (see also Appendix 1).

Primer (DNA or RNA): A short piece of DNA or RNA that directs the site of DNA synthesis. RNA primers occur naturally during replication, whilst the process of polymerase chain reaction relies on purchased DNA primers to ensure the correct sequence is copied.

Prion: A self-replicating protein.

Prokaryote: A simple cell that does not contain a nucleus or other organelles.

Promoter: A part that encodes for gene transcription (production of messenger RNA). A promoter will often include regulatory binding sites for repressors or activators (also known as **operators**).

Protein coding sequence: DNA sequence, beginning with a start codon and ending with a stop codon, which encodes for a protein of interest such as a repressor or an enzyme. See also **open reading frame**.

Receiver: A device that responds to a signal from another cell by transcribing one or more genes as part of coordinating population level behaviour.

Regulatory regions: Regions of DNA that do not encode a protein but instead contain information that directs when, where and how much of a protein needs to be produced.

Replication: The process of copying DNA.

Reporter: A protein which is used as a biological readout by producing a visual output such as fluorescent or colour.

Repressilator: Oscillator constructed from an odd number of inverters coupled together.

Repressor: A protein which binds to DNA in the region of a promoter and blocks transcription.

Restriction enzymes: Proteins purified from bacteria that recognise specific DNA sequences and cut the DNA backbone at these sites.

Ribosome: Cellular machinery for protein synthesis that catalyses the formation of the actual peptide bond.

Ribosome binding site (RBS): Also known as Shine–Delgarno sequence in prokaryotes or Kozak consensus sequence in eukaryotes. A messenger RNA sequence that is recognised by ribosomes and serves to initiate protein synthesis.

RNA: Ribonucleic acid. A biological macromolecule consisting of a sugar-phosphate backbone and four bases: Adenine, Uracil, Cytosine and Guanine. RNA has many functions in a cell. Messenger RNA (see also **mRNA**) is the product of transcription and carries the information to produce proteins within a cell. Transfer RNA (see also **tRNA**) carries amino acids to the correct site for protein synthesis. There are also small RNA molecules (microRNAs) that have been discovered to control cellular processes.

Robotic automation: The use of programmable mechanics, such as robot arms, to automate routine laboratory experiments.

Sanger sequencing: A method to sequence DNA developed partly by Fred Sanger in 1977. The method relies on early termination of DNA replication *in vitro*, followed by electrophoresis on gels to resolve different sized DNA fragments that are used to infer the sequence (see also Appendix 1).

Scale-up: The act of moving a manufacturing process from the laboratory into larger reactors.

Scar sequences: Short DNA sequences that remain at part junctions when two DNA biological parts are assembled together using certain techniques.

Second and third generation sequencing: New technologies for DNA sequencing that have replaced Sanger sequencing. Several of these methods involve hundreds of thousands of reactions occurring in parallel on specialised chips.

Sender: A device that synthesises and secretes a chemical signal that is sent to other cells in a population in order to coordinate behaviour.

Sense (strand): The strand of DNA that contains an open reading frame.

Shotgun sequencing: A strategy for sequencing projects where large regions of DNA are broken into many small overlapping regions that can easily be sequenced and then pieced back together.

Solid-phase oligonucleotide synthesis: A chemical method to custom-make single-stranded DNA pieces fewer than 200 bases long.

Standard/standardisation: Agreed upon measures or designs for parts that can be used by the whole community so that parts are interchangeable.

Statistical moments: Loosely speaking, a quantitative measure of the shape of a set of data points. The most used statistical moments are the first two moments called the mean and the variance respectively.

Stochastic differential equation: Differential equation in which one or more of the terms is a stochastic process, thus resulting in a solution which is itself a stochastic process. Typically, SDEs incorporate white noise which can be thought of as the derivative of Brownian motion (or the Wiener process); other types of random fluctuations are also possible, such as jump processes.

Stochastic model: Mathematical model involving random variable(s), i.e. in which outcomes are not uniquely determined through a known relationship among states and events. Stochastic models are typically used to estimate probability distributions of potential outcomes.

Switch: A triggerable device that alters the behaviour of cells in response to a signal.

Synthetic life: Self-replicating living systems that have been engineered from first principles.

System: A system is a collection of devices that lead to a desired (usually useful) behaviour.

Systems biology: Using mathematical analysis of large experimental datasets to uncover and explain biological function.

Terminator: Also known as transcriptional terminator. DNA sequence, usually with strong secondary structure, that serves to terminate transcription by causing the RNA polymerase to disengage from the DNA strand.

Toggle switch: A particular type of genetic switch composed of two promoter/repressor pairs in feedback with each other.

Top-down: An engineering approach which seeks to first understand high level system behaviour and then decompose systems piece by piece adding complexity where necessary.

Transcription: The process of synthesising an mRNA molecule from a DNA template.

Transfer RNA (tRNA): An RNA molecule with an amino acid attached that participates in protein synthesis by bringing the amino acid to the correct location. This is accomplished through the use of an anticodon which pairs with the codon on the mRNA strand.

Transistor–transistor logic (TTL) data book: Catalogue of standardised parts and circuits useful for electronic circuit design.

Translation: The process of synthesising a protein from an mRNA.

Transposon: A 'movable' element of DNA that is able to excise itself from a strand and move to another location where it can then integrate.

Upstream (DNA): DNA bases closer to the 5' end of the DNA strand than the point of reference.

Index